Published by degrees138 :: degrees138.com

°138

To my Agatha, by whose presence my whole world lights up. She is my sun, my moon, my stars. And a damn fine cook.

To my Caz: Our days together may be numbered, but you will always be my beloved Caz. You have helped me get through some tricky and difficult times, and I would no more countenance not having you in my life than I would countenance chopping off a limb.

It is not the strongest of the species that survives, nor the most intelligent, but those most adaptive to change.
Charles Darwin

———

"Increasingly, your reputation is what Google says it is."
Allan Jenkins, communication analyst and consultant.

ABOUT THE AUTHOR

LEE HOPKINS is a management psychologist and business communicator with nearly 30 years of experience in helping businesses communicate better for improved results and financial returns.

At the leading edge of online business communication in Australia, Lee understands the transformative nature of social media and he spends a considerable amount of time advising businesses, business communities and individual business communicators on the tectonic cultural shifts that new communications technology is facilitating, and how they can best position themselves to take advantage of them.

An internationally sought-after speaker, Lee combines his passion for employee and online business communication with his dynamic presentation skills to create 'once seen, never forgotten' live experiences.

He has written over 300 articles on business communication and some of his many blogposts, podcasts and vidcasts can be found at **leehopkins.com/business-writing**

Lee graduated in 1998 from the University of Surrey with an Honours in Applied Psychology and Sociology. He practiced in the UK as a management psychologist, and co-wrote a number of academic papers, garnered over 1,200 citations, and those citations keep coming.

But the siren song of the internet—and social media in particular—called him and he left management psychology and jumped head-long into website copywriting and the nascent social media world. As a result, he wrote and ran many workshops on the business use of social media, and consulted in Doha, Dubai, San Franciso, Seattle, New Zealand and all around Australia many times.

Along the way, he wrote a white paper and five books on how businesses can co-exist with the anarchic nature of early social media, and do so for the business's benefit. Along the way, he garnered many awards.

In 2018, Lee achieved a Masters degree in Counselling Practice, then practiced for a few years.

He is currently reading and writing for a Masters degree in Creative Writing.

facebook.com/bettercommunicationresults

x.com/leehopkins

instagram.com/bettercommunicationresults

CHAPTER 1
EXECUTIVE SUMMARY

IN THE EVOLVING landscape of social media, businesses face the challenge of not just being present, but dynamically engaging with their audience.

This second edition, updated for 2024, is tailored to guide businesses through the vast options available in the digital arena. From leveraging your website beyond a mere business card to utilising platforms such as Twitter/X, Instagram, Facebook Pages, TikTok, Pinterest, and YouTube, this book provides a comprehensive pathway. It's not just about 'what it is' but 'how to maximise its potential'.

Focused segments on blogging, podcasting, and other digital strategies are designed to amplify your brand's voice and drive traffic directly to your organisation.

Key insights include a referenced article from Neville Hobson addressing the challenges of hate speech on platform X, asserting the ethical considerations brands must make in today's digital climate.

Furthermore, an exclusive chapter will feature a transcript of an insightful interview with Shel Holtz and ChatGPT, offering profound industry perspectives.

The importance of modernising the press release for social media, along with Donna Papacosta's expert breakdown on podcasting across three dedicated chapters, aims to equip businesses with the knowledge to not only participate but lead in the digital conversation.

This edition serves as a utilitarian guide, authoritative in its approach, to navigating and maximising social media's potential for businesses in 2024.

CHAPTER 2
THE BIRTH OF SOCIAL MEDIA

THE ONLINE COMMUNICATION landscape that is Social Media didn't arrive fully formed and ready-to-go—it has enjoyed many years of gestation and training.

Many commentators point to the development of ArpaNet as the birth of Social Media's underpinning technology, but I believe it stretches back further than that.

James Harkin in his brilliant book *Cyburbia* argues convincingly that the online world as we know it has its roots buried deep behind the front lines of World War II, when British military minds were desperately attempting to find ways of accelerating the tracking and targeting of German bombers by those whose thankless task it was to manoeuvre slow and cumbersome anti-aircraft machinery.

One could take the birth of Social Media back even further in time —to an era before computers ('BC'?) and a period when men and women with ideas could stand on boxes and proselytize their views to sometimes disparaging audiences.

Consider, if you will, Judea two thousand years ago—soothsayers, prophets and political 'wannabes' would stand and deliver their views at the risk of boring any audience they could capture, or inciting them to some sort of action.

It's not hard to recall Monty Python's movie *The Life of Brian* in such moments.

But I believe that to find the birthing ground of Social Media we need to look further back in human history, to a time when thoughts and ideas were unable to be written down.

Something like 7.34pm on a mild Tuesday in mid-April, around 165,000 BCE (Before Common Era).

Give or take 50,000 years.

MITOCHONDRIAL EVE

Mitochondrial Eve, the mother of us all, lived on one of the many plains of Africa and with her fellow clan members gathered and shared child-care duties whilst the men hunted.

Come evening they would all gather together and share the fruits of their labours.

The men would recount how they combined forces to track and kill that night's dinner, while Eve and her sisters would discuss local foraging opportunities.

There was no formal language as we would recognise it, but because of a genetic mutation around 30,000 years prior, Eve and her clan *did* have the ability to communicate via speech, albeit with a vocabulary far smaller than ours, and with a slower delivery and simpler grammatical structure; Eve wasn't quite ready to dictate *Hamlet*.

Having progressed past the rock-banging and 'jumping up and down' method of communication, Eve and her contemporaries were sophisticated enough in their communication to be able to manage moving out of Africa and across to Asia, then on to New Guinea, Australia and eventually into Europe.

It is therefore not too big a leap of the imagination to envisage Eve and her fellow clan members discussing, arguing and negotiating about the big and small issues of each day. And thus was a social communication environment born.

Their 'medium' was, of course, primarily speech, possibly augmented with sticks and lines drawn in the sand.

. . .

Media Vs Medium

'Media', according to Dictionary.com, is the plural of 'medium', which they also hold to be—amongst many things—"an instrument or means by which something is conveyed or accomplished" and "one of the means or channels of general communication, information, or entertainment in society."

My trusty and dog-eared New Collins Concise English Dictionary agrees that 'media' is the plural of 'medium', which it holds, inter alia, to be "an intervening substance or agency for transmitting or producing an effect; vehicle" and "a means or agency for communicating or diffusing information, etc., to the public", further noting that "careful writers and speakers do not use *media* as a singular noun when referring to a medium of mass communication: 'television is a valuable medium (not media) for advertising."

We have been warned.

Thus 'Social Media' can arguably be various channels or instruments by which ideas can be expressed, shared, debated and/or negotiated.

Looking at the development of human communication over the millennia, I believe that various channels or vehicles for expressing, sharing, discussing and negotiating have come about as homo-sapiens like Mitochondrial Eve progressed their tribe along the path of civilization.

It is the developmental work of only a few hundred thousand years from speech-rendered proto-language to the sophisticated word and visual play of today's digital multimedia artists—and note that we have five times more words to play with than Shakespeare had in his time.

So let us hop in that handy time-machine over in the corner of the cave and fast-forward to 2004—arguably the year that everything started (or 1sm).

Actually, the technological seeds of 2004 are rooted further back, to 1999 and beyond, but for the sake of simplicity and illumination let's focus on 2004.

. . .

Blogging

In 2004 'blogging' came to the world's attention; personal online publishing moved out of the hitherto 'IT geek' domain and into the consciousness of the general public. Folks *other* than IT evangelists started using websites such as blogger.com, livejournal.com and type-pad.com to publish their personal news and views on the issues that affected them.

The Wall Street Journal ran an article in 2007 (August 15) wherein the author claimed that the blog as a distinct life-form started on December 23, 1997 with the publication by Jorn Barger of his site, *Robot Wisdom*.

That claim was quickly shot down by the blogosphere (something that the blogosphere is exceptionally quick and able to do, as has been noted by chagrined politicians), who pointed out that Steve Jackson had been publishing a site called *The Daily Illuminator* since December 1994.

Such a heady claim was subsequently trumped by EFF (Electronic Frontier Foundation) chairman Brad Templeton, who speculated that the 'blog' has its origins pre World Wide Web, residing in the moderated newsgroup *mod.ber*, run by found Brian E. Redman (the *ber*) and friends. Templeton went on to claim that his own moderated newsgroup, *rec.humor.funny* / netfunny.com, may possibly be the longest still-running blog, being nearly 22 years old, beginning life on 7th August 1987.

What ALL of these sites allowed was a serial publishing schedule, like a journal, written as a series of new items on a semi-regular basis, with a coherent and personal editorial voice (whether that 'voice' was the voice of one person or a small team) and the ability for readers to comment in some way.

Podcasting

Also in 2004 the Apple iPod became the mp3 player of choice for the 'digiliterati' (digital literati) and the cashed-up. It seemed that everyone who had one gloated about it and those that didn't have one envied those who did.

So iconic became the iPod that in some communities—technology, Silicon Valley, *inter alia*—you weren't taken seriously by your peers unless you had one.

Various technologies already existed to enable you to transfer music from your computer to your iPod, but when online audio pioneer and former MTV presenter Adam Curry (aka 'The Podfather') and coding guru Dave Winer joined forces things really accelerated. They wrote a small piece of code that allowed for attachments to be sent via RSS feed to users' computers and thus the Era of Podcasting began.

Podcasting is many things to many people—"radio with a rewind button" is my personal favourite—but no matter what one's choice of material (music, voice, sound effects, experimental sound theatre, walking soundscapes, and so on) the technology allowed the individual yet another channel with which to broadcast their views, engage in inter-podcast discussions with other podcasters (broadcaster-to-broadcaster dialogue), provide thought leadership for their community and potentially find new audiences with which to engage.

For the business communicator, January 4, 2005 saw the first release of *For Immediate Release*, a then-weekly now twice-weekly conversation between two highly experienced business communicators—Shel Holtz and Neville Hobson—and their community.

For Immediate Release spawned a host of excellent business/PR/marketing podcasts, but remains the 'Daddy Mac' of them all.

There is a list of useful podcasts at the end of this chapter.

When Apple upgraded an early version of its iPod music management and transfer program, *iTunes*, to allow the capture and transfer of podcasts, and at the same time allowed podcasters to list their podcasts as available for download via *iTunes*, the iPod-owning audience suddenly had their ears opened to a whole new range of material.

From music bands giving away their music for free, to voice actors giving audio books away as a showcase for their voice talents, from devotees rebroadcasting 1930s and 1940s radio shows to universities rebroadcasting lectures... suddenly an iPod was no longer an expensive portable music player but a portable university, radio station, gossip columnist and industry news provider, all in one sleek and desirable package. Overnight *iTunes* became the leading music transfer software, putting quite a few others out of business, and the iPod continued to steamroll over any competition and completely dominate the portable music/video player market.

Video

With the advent of cheap video production, either via webcam or the rise of cheap digital camcorders (and still digital cameras with movie capabilities), came the rise of YouTube and other video sharing sites.

Along with this rise came the ability to subscribe to video podcasts via *iTunes* and watch them on a computer monitor, a video iPod or a mobile phone.

Of course, such features have come with a plethora of new names, names which can cause confusion.

For example, no general consensus has yet been reached on what to call video podcasts—*vodcasts, vidcasts, vidblogs, vlogs* have all been used to varying degrees in various geographic regions and with varying success.

Video sharing sites

There are lots of video-sharing sites on the internet, the top 10 sites arguably being:
· YouTube
· MetaCafe
· Break
· Google Video
· DailyMotion
· Yahoo! Video
· Revver
· Vimeo

· vidLife
· Stickam

There are also peer-to-peer video conversation sites such as ooVoo and Seesmic that offer the ability to engage in video conferencing via simple webcam and microphone (usually a simple headset microphone will suffice in terms of quality). The VOIP (Voice Over Internet Protocol) service providers Skype and Zoom also allow for video conferencing.

These sorts of sites aren't geared for (or even 'about') high production values, but they are about using video to take multi-location audio conversations to the next level.

Even the term *podcast* itself continues to cause confusion.

Contrary to initial public perception, you *don't* need an iPod to listen to a podcast. In fact, you can use any mp3 player, including computer-based ones, to listen to or watch a podcast or vidcast (the term I have settled on for video podcasts).

The confusion over 'podcast' resulted in innumerable questions being asked in 'awareness and training sessions' that many of us in the Social Media consulting industry ran (and still run—and yes, we still get asked if you need an iPod to listen to a podcast).

KEEPING THE CONVERSATION FLOWING AROUND THE WEB

But whether the channel of communication is a blog, or a podcast, a vidcast or even one of the newer tools such as a wiki or Twitter (see *Chapter 5: Tools* for a more in-depth discussion of these two tools) the fundamentals remain: online communication in a 'human' voice with the ability for others to respond—either directly, or indirectly via their own preferred channel and with a clickable link back to the original communication.

A human voice

"To have a conversation, you have to be comfortable being human—acknowledging you don't have all the answers, being eager to learn from someone else and to build new ideas together.

"You can only have a conversation if you're not afraid to be wrong. Otherwise, you're not conversing, you're just declaiming, speechifying, or reading what's on the PowerPoints. To converse, you have to be willing to be wrong in front of another person.

"Conversations occur only between equals. The time your boss's boss asked you at a meeting about your project's deadline was not a conversation. The time you sat with your boss's boss for an hour in the Polynesian-themed bar while on a business trip and you really talked, got past the corporate bullshit, told each other the truth about the dangers ahead, and ended up talking about your kids—that maybe was a conversation."

The Cluetrain Manifesto, p.123

The clickable link back to the original communication is key to keeping the conversation flowing around the web. The ability for anyone to come along, pick up a conversation half way through and track it back to its antecedents is a vital component of social media.

In this way it doesn't matter whether a comment is left on, for example, the original blog post or is left on the commenter's own site. Indeed, it doesn't even matter what channel is used to reply. One could, for example, leave a comment to a blog post in the form of a video, or an audio comment. Person B can comment about Person A over on Person C's website.

Or Person B can leave an audio comment about Person A's video to be played in Person D's podcast.

As long as a way is given for others to access the 'start' of the conversation then the conversational thread can be followed by others.

Link love

All of this cross-linking (aka 'link love') serves several purposes, some more altruistic than others.

Naturally, it is good manners to reference the source of the original material; that way anyone reading, listening or watching a comment can quickly research and understand the context of the comment.

In addition, it is handy for the originators of content to know when they have been talked about. Thankfully, there are several mechanisms by which an author can find out who is talking about them.

> "There is only one thing worse than being talked about... and that's NOT being talked about!"
> Oscar Wilde

The ability to see who is 'talking' about you in a fundamental part of managing one's online, digital reputation; with every comment—be it brickbat or bouquet—now being captured digitally and stored forever, the need to monitor what is being said about a company, its products and services, and its key personnel is finally being understood by the business community.

> "Increasingly, your reputation is what Google says it is"
> Allan Jenkins, communication analyst and consultant

We will look at 'monitoring' in further detail in *Chapter 3: Social Marketing*, but in the meantime don't underestimate the importance of knowing what is being said by others. It can sometimes be dismissed as 'ego-surfing', but finding out who has linked back to what content, and who has mentioned it but not linked back, is vitally important to any corporate (and personal) social media success.

Wikis

If the social media community itself is undecided about what to call video blogs, then the general business community is equally undecided about how to define what a wiki is and what use it might have 'inside the firewall'.

Ward Cunningham coined the term 'wiki' back in 1994, basing it on the Hawaiian word for 'quick'. Since then it has become a firm favourite in the toolbox of many Knowledge Management gurus and tech-savvy project managers, but alas relatively unappreciated across the wider business community.

The most well-known wiki is, of course, Wikipedia—the online reference site that is currently five times the size of the Encyclopaedia Britannica and, according to various academic and independent studies, equally as accurate.

Wikis can have a tremendous impact on a company's knowledge bank. For example, some employees at British Telecom ('BT' as it is known) launched a small wiki with no fanfare or publicity. Curiosity and word-of-mouth grew it in size and popularity to levels far in excess of the developers' dreams, and at a speed that shocked them (see *Case Study: BT's wiki*).

A wiki is a content management system (CMS) that is open to alteration, addition and amendment by more than one person. Wikis bypass the traditional corporate roadblock—email—to enable rapid collaboration and information sharing.

My namesake, Lee LeFever, and his wife Sachi have created a fabulous series of videos that explain much of the various elements of social media in very simple terms, including the wiki. I strongly advise you to visit CommonCraft.com and watch the videos yourself, then share them with your teams. There are even high-resolution versions available for purchase with distribution licences.

But, in essence, a wiki allows a project team to agree a common set of documents, terms, details, and so on, without the endless to-ing and fro-ing of emails, and without the rapidly confusing visual mess that a Word document descends into after more than two alterations by two different collaborators.

Wikis solve office politics problems

Wikis also help get office politics out in the open and dealt with so that the *real* work can get done.

Consider this: your project entails input from a formally-titled business communicator, a lawyer, a project manager, a marketer, an inventory manager, a product manager and a financial gatekeeper.

Now, your experience may differ ("your mileage may vary" as my North American colleagues delightfully say), but in my 25-plus years of business communication I have yet to meet a lawyer who didn't think that *their* command of the spoken and/or written word was superior to most (if not all) of the other project team members. Thus their belief that *theirs* should be the final wording, as it is the clearest, most concise and logical, whilst still defending the company against any risk.

Equally, I have yet to meet *any* member of a team who didn't believe that *their* principal concern was not worthy of a Top Three spot of concerns that the project *must* address.

So every time a draft document was circulated for comment, back would come conflicting viewpoints and emphases.

Document collator/editor

Lawyer

Accountant

Marketer

Manufacturing/ Production Manager

Sales

Customer Service/Support

Product Manager

Inventory Manager

!*!@!

Having spent far too many years lost down the rabbit hole of piecing together paragraphs of conflicting information and dealing with resultingly upset egos, I can assure you that a wiki allows the egotists to 'fight it out' in public as documents and pages go through their various iterations. If one particular team member becomes too insistent on their point of view, despite comments and revisions from others, then the team can point this out to them, an action that usually very quickly modifies over-exuberant behaviour.

Should the team member continue to insist on their point of view being the dominant, resulting in unnecessary delays to the project and increasing disharmony, the project manager can have a quiet word in their ear, or else escalate the issue higher up the management chain for adjudication.

Irrespective of harmony or disharmony, the group always works from the one webpage to shape and agree on a final version. Each section or sub-section of a document, report or project can be its own standalone webpage, editable by only one person at a time (to avoid any data conflicts) and with a toolbar that offers the standard word processing options such as bold, italic, heading and sub-heading, add a link, add an image, and so on.

When a page has been agreed and finalised, a simple copy-and-paste can take it from the wiki into a more formal document created within a word processor or desktop publishing application. Any stylistic variations can be accommodated by the style sheet underpinning the word processor or document layout tool (Word, Illustrator, etc.).

Wikis have taken off within the large enterprise, particularly amongst knowledge management teams, who leverage the power of the wiki in order to shorten project design and delivery time.

RSS

A tiny piece of computer code allows for unparalleled knowledge sharing. That code has been called 'rss', although exactly what rss stands for is sometimes the subject of debate.

'Really Simple Syndication' is arguably the most accepted definition, highlighting how content from one content provider can be syndicated or re-published on the web property of others. In this way, for example, anything I write on my blog can be automatically inserted into someone else's webpage.

This is how news sites like BBC.com and ABC.net.au are able to fill their home pages with constantly updating and changing headlines; rather than have a web team of sub-editors constantly re-writing and editing webpages, rss code 'pulls in' data from the company's servers, wherein the style sheet (a piece of computer code that web browsers read in order to know how to display the content in your browser) formats the 'look and feel' and placement of the data.

Rss allows for the separation of content from presentation, allowing for a single piece of data to be delivered across multiple platforms and formats, and in multiple documents for multiple uses.

The data itself can be shared and republished both in front of and behind the company firewall, so that the public can access some of the data and employees other data, perhaps with more in-depth content or company-sensitive analysis.

There has been a tremendous growth in the use of rss behind the firewall. Fuelled by employees who routinely use this technology outside of work and often from their work computer, many organisations have introduced personalised employee home pages where the data presented to the employee once they have logged into the network is content delivered automatically from pre-determined content providers deemed useful for that particular employee in their particular department or work environment.

Additionally, many organisations also allow their employees to search, find and subscribe to content from other, not necessarily related content providers, both inside and outside of the organisation. In this way employees can stay abreast of developments in their industry, movements by their competitors, and keep in touch with thought leaders from their own and allied professions.

Some of the earliest adopters of social media into the organisational communication matrix took particular care to 'not scare the natives': they don't use the 'buzzword' terminology of social media, such as rss, feeds, blog, podcast, wiki, and so on, but instead choose to use neutral or old-fashioned terms for the new tools. They also offer multiple ways for employees to access or subscribe to these data sources, including good ol' email.

After all, it is not the tool itself that is transformatory within an organisation, but how it is used. If employees are comfortable using email technology then there is little point forcing them to change their information-receiving style. The less resistance to change, the smoother the cultural transformation and the quicker the improved information process can start delivering its potential.

Micro-blogging (aka Twitter)

If one technology has taken off to the amazement of even us 'wizened' old social media salts, it is Twitter (now known as X).

Since entrepreneur Elon Musk took it over, X has lost billions of dollars and become the default home of right-wing hate groups, as well as conspiracists and the anti-science brigade.

Trying to explain in simple-to-understand terms what X is... well, let's just say that herding cats would be an easier task!

At one extreme it is akin to a text messaging service, allowing anyone with a free Twitter account up to 280 characters at a time to 'say something'. Every time a message is sent (a 'tweet' as the 'Twitterati' call it; don't worry, the (ab)use of the English language with 'twit' as a prefix knows no bounds and shows no sign of stopping), it is stored in a database, wherein it can be searched by Google and other search robots, including Twitter's own. Each person's message stream is rss-enabled, allowing anyone to subscribe to their feed.

These rss feeds can either be read in traditional rss readers (such as Google Reader (sadly, Google shut down this brilliant tool a few years ago), FeedDemon, GreatNews, *inter alia*) or in a dedicated tools such as TweetDeck or Twhirl (the two current favourites) for your desktop/laptop, or ceTwit, Jitter, Hahlo, or PocketTweets amongst many for your smart phone/iPhone.

Why would anyone DO that?

Upon first seeing Twitter I was not alone in asking, "Why would anyone DO that?" However, it only took a short while to become enamoured with it.

Here are just some of the many uses to which Twitter/X can be put:

- Comcast, Network Solutions, JetBlue, H&R Block and Zappos use Twitter as a way of creating and sustaining customer loyalty and satisfaction;
- Dell uses it as an 'early warning radar' for customer service issues that need attending to before the customer becomes 'heated';
- A local public radio station in Melbourne, Australia, used Twitter to keep the public aware of what was happening during the disastrous Victorian bushfires;
- Californians are kept similarly informed during their own fire seasons;

- Mzinga uses Twitter to offer user promotions and to enter into real conversations with its customers;
- British Telecom, having purchased two major businesses in the security industry, uses Twitter to engage with the security industry about future trends;
- Evernote used Twitter to invite people into a private beta of their software, continuing to use Twitter as a bug reporting and feedback channel;
- General Motors uses Twitter to connect with its customers. 'We usually try to Tweet some sort of question for our followers every day. These questions can be vehicle-related (i.e. What's your favorite convertible?) or just for fun (i.e. What's your favorite roadside diner). This two-way approach continues to work well for us and we've seen a number of people tweeting more about GM because of our presence there,' says Adam Denison, GM Social Media Communications;
- EMC uses Twitter to not only push out press releases about new products and services, but has also created a separate Twitter profile in order to communicate to participants at their conferences. EMC use it to send people to different on-site podcasts, blog posts by attendees and Flickr photos, as well as direct people to contest areas, keynote speeches, and so on;
- The Home Depot uses Twitter to promote the 'out-of-store' lives of their employees, helping create a 'family' feel to the store;
- Baskin-Robbins use Twitter to promote 'feel-good' offers such as this one, 'Join us for 31 Cent Scoop Night at Baskin-Robbins and help us honor America's firefighters. Participating stores will reduce prices of small ice cream scoops to 31 cents';
- Publishers John Wiley & Sons have several employees who use Twitter to find new authors, talk with existing authors and customers and use it for market research: 'Would you

buy a book on X?` and 'What cover would make you more likely to buy this book?`;

- FreshBooks uses Twitter to handle small support issues, update friends and fans of new developments, and listen to its customers;
- Plaxo use Twitter to engage with its members and potential members, as well use Twitter's search capabilities to see what is being said about Plaxo in the Twittersphere. That way, says John McCrea, Plaxo's head of marketing, he can reach out to answer questions, help solve problems, and correct any misperceptions;
- Upon first seeing Twitter I was not alone in asking, 'Why would anyone DO that?` However, it only took a short while to become enamoured with it. Here are just some of the many uses to which Twitter can be put:
- Advertising a business;
- Appointment reminders;
- Complaining about a business;
- Complaining about a job;
- Crowdsourcing for resources;
- Disaster alerting and response;
- Emergency response team management;
- Environmental alerts: pollen counts, pollution levels, heat waves, severe weather alerts;
- Exercise management and encouragement;
- Getting feedback;
- Hazardous materials communication;
- Issuing alerts for missing nursing home residents;
- Issuing asthma alerts;
- Launching a business;
- Live coverage;
- Looking for a job;
- Networking for a job;
- Notifying customers;
- Office staff to patients about appointment reminders;
- Personal branding;

- Physician to Physician communication for general medical questions and 'curbside consults';
- Physician-to-team member about non-urgent matters;
- Posting a job;
- Psychiatric 'check-ins' for patients;
- Advertising your blog;
- Networking with friends;
- Finding like-minded people;
- Debating;
- Making 'To-do' lists;
- Tweet a poll or survey;
- Following or contributing to election campaigns;
- Conducting structured interviews;
- Showcasing useful product 'how-to' videos you have released on YouTube;
- Providing or requesting customer support;
- Searching for industry contacts, other industry members and industry news via http://search.twitter.com;
- Customer/supplier Q&A sessions;
- Brainstorming amongst your customers, clients, suppliers and evangelists;
- Providing weather forecasts for events you are running;
- Running promotions and contests;
- Alerting colleagues of changing flight times;
- Tracking your company's standing on the US Stock Exchange: stocktwits.com;
- Offering BETAs of your company's software;
- Outsourcing tasks;
- Reading breaking news;
- Setting up meetings; Taking notes;
- Tracking disease-specific trends;
- Updates/changes at events (e.g., speakers,venues); and
- Watering your plants - you can buy a kit that hooks up to the soil of a plant and have it tweet when it needs watering or is overwatered!

Still confused? Watch this video: https://rebrand.ly/cddda1

Conclusion

We are just at the start of our journey across this new communication landscape, but already we have briefly met many of the most-used tools and platforms.

But lest you think that the Social Media world is confined to the usual flat, two-dimensional, toilet-roll type pages of the web as we have known it until now, be assured that there is a whole other web out there, a web full of colour, movement, interaction, people and places.

It is a web of monsters and aliens, of butterflies and angels, of werewolves and vampires, of 16th century royalty and 22nd century cyborgs.

It is a web run entirely by people like you and I, and you just need to turn the page to begin your labrynth-like descent into 2.5D and 3D web worlds that reflect all of human creativity, ingenuity and madness back at us.

See you there...

CHAPTER 3
IBM SOCIAL COMPUTING GUIDELINES

1. Know and follow IBM's Business Conduct Guidelines.
2. IBMers are personally responsible for the content they publish on blogs, wikis or any other form of user-generated media. Be mindful that what you publish will be public for a long time—protect your privacy.
3. Identify yourself—name and, when relevant, role at IBM—when you discuss IBM or IBM- related matters. And write in the first person. You must make it clear that you are speaking for yourself and not on behalf of IBM.
4. If you publish content to any website outside of IBM and it has something to do with work you do or subjects associated with IBM, use a disclaimer such as this: "The postings on this site are my own and don't necessarily represent IBM's positions, strategies or opinions."
5. Respect copyright, fair use and financial disclosure laws.
6. Don't provide IBM's or another's confidential or other proprietary information. Ask permission to publish or report on conversations that are meant to be private or internal to IBM.

7. Don't cite or reference clients, partners or suppliers without their approval. When you do make a reference, where possible link back to the source.

8. Respect your audience. Don't use ethnic slurs, personal insults, obscenity, or engage in any conduct that would not be acceptable in IBM's workplace. You should also show proper consideration for others' privacy and for topics that may be considered objectionable or inflammatory—such as politics and religion.

9. Find out who else is blogging or publishing on the topic, and cite them.

10. Be aware of your association with IBM in online social networks. If you identify yourself as an IBMer, ensure your profile and related content is consistent with how you wish to present yourself with colleagues and clients.

11. Don't pick fights, be the first to correct your own mistakes, and don't alter previous posts without indicating that you have done so.

12. Try to add value. Provide worthwhile information and perspective. IBM's brand is best represented by its people and what you publish may reflect on IBM's brand.

CHAPTER 4
BT'S WIKI

THREE COMMUNICATION- and IT-savvy folks at BT (British Telecom) decided to work a small number of hours after work for a few days (about three, I believe) and knock together a wiki.

Released unannounced and unpublicised, with just a small logo in the corner of BT's default intranet as a link to it, "BTpedia: BT's Collaborative Encyclopedia" had a very slow beginning and no training or instruction as to what it was there and how employees were to use it.

That was part of a deliberate strategy by the Head of Communication Services at BT, Ross Chestney, who wanted to see what would happen next.

Slowly at first, but with accelerating pace, BTpedia took off to the point where it now has more visitors and content than BT's 'traditional' intranet.

Ross and BT Internal Programme Manager Richard Dennison are also using project-based wikis to enable greater cross-collaboration and build communities of interest, as well as continue to build the company's shareable knowledge base.

Ross and Richard have a great slideshow presentation that highlights all of BT's various social media initiatives and the lessons they have learnt over at slideshare:

http://www.slideshare.net/whatidiscover/social-software-in-a-corporate-context-presentation

CHAPTER 5
MANAGING TIME WASTING AND VIRTUAL RABBIT HOLES

EMPLOYERS WORRIED that their employees will waste endless hours accessing non-work material already have access to two valuable tools to manage the risk:

- KPIs (Key Performance Indicators); and
- Shame.

Individual productivity is still governed by agreed job performance requirements and KPIs.

If employees are frequenting social networking sites at the expense of their performance, it usually only takes a quiet word in their ear from their manager to bring them back into line. If this fails, the usual disciplinary measures can be utilised.

Additionally, nothing spreads around a company faster than gossip, and firing an employee for gross negligence, dereliction of duty or grossly-inappropriate behaviour, will generate a non-verbal message that will quickly rip through the hallways if unofficially allowed.

Add to it a memo from HR to the whole company that explains why the unnamed employee was dismissed and you have introduced a prophylactic against further abuse.

In this current business climate few can afford to lose their job, let alone their reputation and future employability owing to an act of extreme foolishness, and an attitude of 'they're not smart enough to catch me' or 'it will never happen to me'.

NAVIGATING TIME WASTING AND DIGITAL RABBIT HOLES ON SOCIAL MEDIA PLATFORMS

In an era where digital platforms serve as the primary conduits for both information and entertainment, it's increasingly challenging to manage the siren call of social media. The allure of the endless scroll, the magnetic pull of doomscrolling—these are but a few examples of time-wasting behaviours that not only interrupt productivity but can also encroach on one's mental well-being.

Understanding Time Wasting on Social Media

Social media time-wasting isn't limited to the occasional divergence from daily tasks; it encompasses behaviours like endless scrolling and doomscrolling through feeds that are seemingly inexhaustible. **Psychologist and social media expert, Dr. Linda Kaye**, explains that "the design of these platforms leverages powerful psychological principles such as the variable reward model, which hooks users into cyclical usage patterns."

The Rise of Digital Rabbit Holes

Digital rabbit holes, such as those found on Facebook, YouTube, Instagram, and TikTok, are enticing trails of content that capture the user's attention and tap into the natural human propensity for curiosity. Sadly, the descent into such rabbit holes can often lead to the consumption of content steeped in negativity with severe implications for one's disposition.

Tristan Harris, a former Google Design Ethicist and co-founder of the Center for Humane Technology, cautions about intentional designs in these platforms that amplify incendiary and sensational content to increase user engagement, often at the cost of the user's peace of mind.

Management Strategies to Combat Time Wasting

For individuals looking to combat the time lost to digital platforms, sound management strategies are essential. Management consultant **Michael K.**, advocates setting definitive goals and actionable priorities. "Without clear objectives, our attention defaults to the most immediately engaging stimuli," he notes.

Utilising time management methods—such as the Ivy Lee method, which promotes tackling the most important tasks first—can provide structure and help mitigate distractibility. Productivity tools and apps also present viable means to methodically manage and monitor use.

Techniques to Avoid Social Media Rabbit Holes

Avoiding these rabbit holes requires conscious effort and purposeful tactics. Cal Newport, author of "Digital Minimalism," suggests, "Intentionality is crucial when interacting with social media. Ask yourself whether this is the highest value activity you could be engaging in."

Adjusting social media feeds to display content that is uplifting and removes negative triggers is also a tactic upheld by Nir Eyal, author of "Indistractable," who describes it as managing your "external triggers to create a more wholesome digital environment."

Impact of Rabbit Holes on Mental Health

The negative effects of falling into digital rabbit holes on mental health cannot be understated. A steady diet of adverse news and distressing content leads to the phenomenon known as 'Vicarious Trauma.' From my history as a psychologist, the continuous exposure to distressing content leads to increased anxiety and stress, which are detrimental to mental health.

Cultivating a strategic approach to protect psychological well-being whilst navigating social media is imperative. Techniques include designating "news-free" times and engaging in digital detoxes, as well as seeking balanced narratives.

Case Studies and Examples

Institutional examples abound wherein commitments to digital mindfulness have reaped dividends. Companies such as **Slack Technologies, Inc.,** have instituted 'no scroll' policies during work hours, ensuring focused productivity. Elsewhere, individuals have documented their journeys to overcome the seduction of digital rabbit holes, transitioning from passive consumers to curated users.

FUTURE TRENDS AND RECOMMENDATIONS

Emerging technologies offer promising techniques to aid in combating digital time wastage. AI-driven apps capable of predicting when users might drift into unproductive usage and offering redirection are currently in development.

Social media platforms themselves are also evolving, incorporating inbuilt limitations to remind users to take breaks. The implementation of AR and VR as methods for professional development presents alternatives to passive scrolling—immersive experiences that educate rather than entertain minimally.

CONCLUSION

This investigation into the battlefield of social media indicates that effective management of our online activities is not merely preferable, it is necessary. The ability to recognise and resist digital rabbit holes, complemented by informed strategies and tools, can liberate us from the digital labyrinth's constraints.

A parting consideration from **Cal Newport**: "You must become deliberate with how you engage with social media—turn off notifications, schedule your access, curate your feeds. This is how you reclaim authority over your time and attention."

In subscribing to these practices, individuals empower themselves to foster a more constructive relationship with social media, one where the instrument of technology serves us, rather than us invariably serving it.

3 TIPS TO REMEMBER

1 Consistency Overcaprice: Fostering judicious social media habits is an exercise in discipline. Regular examination of social media habits fortifies autonomy over one's digital life.

2 Resourcefulness: Employ every available means—be that applications, settings within platforms, or shared tactics—to ensure that the time spent on social media aligns with productive, fulfilling objectives.

3 Acknowledgement of Progress: Registering one's advancement is as crucial as the adherence to the strategies themselves. Each act of resistance against the pull of the digital void is a step towards mastery over one's social media landscape.

CHAPTER 6
SHEL HOLTZ INTERVIEWS CHATGPT

CHATGPT AND SHEL Holtz

I listened to a podcast interview intently yesterday, and was blown away. ChatGPT—and AI in general—is way more advanced than I have given it credit for.

There seems a HUGE amount of dog work that AI can take off the overworked and underappreciated communicator, allowing said communicator to focus on higher, more strategic and longer-term work, including being able to reposition themselves as a resource of advice for senior management, not just that 'person who does that communication stuff, whatever that is'.

I've been using 'Jasper' for nearly a year and until yesterday didn't realise how criminally underused it was in my hands. Already, I have upped my game to create better headlines and subheads, write more interesting chapter titles, and so on.

The interview between Shel and Ms ChatGPT is riveting, and I strongly advise you to listen to the hour long interview and marvel at how much more you could ask your AI engine of choice to do.

Brilliant episode and interview. Kudos, Shel.

———

Volumes have been written in the last year about applying ChatGPT and other generative AI tools to public relations and organizational communication. In this midweek episode of the "For Immediate Release" podcast, we let ChatGPT speak for itself about how it can enhance communicators' work — and whether communicators are at risk of being replaced by AI. I'm solo in this episode, so I interviewed ChatGPT to ask about its processes for press releases and other PR writing and editing, its non-media and non-writing capabilities, ethical considerations, crisis communication functionality, and much more. Hear ChatGPT, in her own words, articulate the best approaches professional communicators can take to optimize their work with an AI collaborator.

https://www.firpodcastnetwork.com/fir-375-chatgpt-speaks-for-itself-about-its-pr-capabilities/

Shel Holtz: If you've been listening to FIR at all, you know what Neville and I think about ChatGPT and generative AI in general. What does ChatGPT think? You'll find out in this midweek episode, but not a short one, of 'For Immediate Release', another fine podcast from the FIR Podcast Network.

Heidi Miller: This is 'For Immediate Release', the podcast for communicators.

* * *

Shel: Hi everybody, and welcome to episode number 375 of For Immediate Release. This is Shel Holtz. And no, there's no Neville this episode. This is the week between Christmas and New Year's, and Neville and Laura are away, and I thought I would take advantage of this downtime to do something I've been wanting to do for a while, and that is conduct an interview with Chat GPT.

Of course, now you're able to engage in a vocal conversation with Chat GPT, so that's what I've done. Neville and I have been talking about AI generative AI in particular for about a year now, and we've been raising a lot of issues, and I've been curious what ChatGPT thinks of all of these matters, so I've taken these questions directly to the source.

Now, I do want to share a couple of notes before I jump into the interview. First, is that this is not an original idea, the idea of having an interview with ChatGPT , I wish it were, I wish I had come up with this on my own. But somebody, I want to say it was Donna Papacosta, alerted me to a faith-based podcast called Nomad, I believe out of the UK, in which one of the hosts interviewed ChatGPT .

Interestingly, this was before it had the voice conversation capability. He used some other mechanism to convert ChatGPT 's answers into a voice. I didn't do that. I actually used the voice functionality that's built into the mobile app. But I did run into one significant challenge. I took the better part of an afternoon trying to resolve it and couldn't.

And that is feeding the audio from the app into an audio device. I tried several different things. I tried Taking the data directly through a cable from the data port on the phone into an audio device and that didn't work. It continued to come out of the speakerphone. Everything else that was internal audio on the phone played through the audio device.

I'm talking about YouTube audio podcasts, music, you name it, it behaved the way it was supposed to, but ChatGPT would only play through the speakerphone. I also tried using a native audio recorder on my phone. Same thing. Everything else recorded the way it was supposed to, even phone calls, but ChatGPT just wanted to play through the speakerphone.

So, what I ended up doing was putting my phone in a bracket on a tripod and setting it right next to the microphone, the very microphone I'm talking to right now. So, the audio that you're going to hear from ChatGPT is not quite as good as I was hoping it would be. It's certainly listenable. But it's clear that you're listening to the speaker phone being recorded through the microphone.

Second thing I wanted to point out is that I gave ChatGPT a considerable list of instructions before we started the interview. I told ChatGPT who I am. I explained the nature of For Immediate Release, the fact that we have been talking about generative AI since ChatGPT 3. 5 was released in November of 2022, and that we are focused on public relations and organizational communications and that I wanted to ask a lot of questions around that and that I was hoping to have a conversation that would sound good and right in a podcast. So, what you're about to hear is the result of that. And afterwards I'll be back with a couple of quick observations.

And by the way I am cross posting this episode of FIR to 'FIR Interviews'. No reason that you should have to listen to both. It's exactly the same file. So here is my interview with ChatGPT .

* * *

Shel: Let's start with some of the fundamentals. You were trained on a lot of various types of documents. When you're prompted to help with a PR or communications task, what type of material from your training set would come into play?

ChatGPT : When I'm asked to assist with a PR or communications task, I utilize a variety of materials from my training set. This includes public relations strategies, communication theories, press releases, media relations content, crisis communication plans, and digital marketing trends. I also draw from real world case studies and articles about successful PR campaigns and industry best practices.

This wide range of sources helps me provide informed and nuanced advice for crafting messages, engaging audiences, and navigating the ever-evolving digital landscape in public relations.

Shel: What are some of the more common uses to which you're being put in support of public relations and communications?

ChatGPT : In public relations and communications, I'm commonly used for several tasks. These include crafting and editing press releases, generating ideas for content, aiding in crisis communication planning, monitoring brand reputation, and suggesting communication strategies.

Shel: A lot of the criticism about using you to write, say, press releases is that that process loses the human element.

But most press releases written by humans don't seem to include much of the humanity that people lament is missing from an AI generated release. Do you think a press release that you generate is good enough?

ChatGPT : You make a valid point about the human element in press releases. While it's true that many press releases tend to be formulaic, and might not always reflect deep personal touch, AI generated content, including press releases, can indeed be quite effective.

I strive to produce clear, concise, and relevant content that aligns with the established norms and formats of press releases. However, the best approach might be a blend of both AI and human effort. AI can draft a press release, ensuring it contains all the necessary information and adheres to the standard structure.

Then, a human can add nuances, personal insights, or company culture elements that reflect the brand's voice and human touch. This collaboration can result in a press release that is not only informative and professional, but also resonates more with the intended audience.

Shel: I talk to a lot of communicators who are using AI tools on an *ad hoc* basis, but not many who have really adapted their workflows to include AI and account for it. How do you think communicators should go about thinking about how AI should affect their workflows?

ChatGPT : Communicators should approach integrating AI into their workflows thoughtfully, focusing on augmentation rather than replacement. Here are a few steps to consider. One, identify repetitive tasks, determine which tasks are time consuming and routine, such as media monitoring, data analysis, or drafting basic content, AI can handle these efficiently freeing up time for strategic thinking and creativity too.

Enhance creativity and strategy. Use AI for brainstorming and generating ideas, but let the final creative and strategic decisions stay human-led, ensuring that the unique human touch and understanding of context are preserved. Set clear objectives. Understand what you want to achieve with AI, whether it's increasing engagement, reaching a wider audience, or improving response times.

This will guide you in selecting the right tools and methods. Educate and train the team. Ensure that your team understands how to use AI tools effectively and ethically. Continuous learning should be encouraged as AI and its capabilities evolve. Five, monitor and adapt. Regularly review the impact of AI on your workflows.

Be open to adjusting your strategy as new tools emerge and as you better understand the strengths and limitations of AI. By gradually integrating AI, focusing on its strengths, and always keeping the unique value of human judgment and creativity at the core, communicators can effectively enhance their workflows and outcomes.

Shel: I want to go back to the press release for a minute. When a public relations practitioner is working with you to create a press release, what kind of information should they include in the prompt in order to help you produce the best possible output?

ChatGPT : When a public relations practitioner is working with me to create a press release, they should include:

1. **Objective.** Specify the goal of the press release, such as an announcement or event details.

2. **Key messages.** Outline the main points that need to be communicated.

3. **Target audience.** Identify who the press release is intended for.

4. **Facts and figures.** Provide relevant data, statistics, or specific facts to include.

5. **Quotes.** Offer quotes from involved parties or stakeholders for authenticity.

6. **Boilerplate information.** Include standard company information and contact details.

7. **Tone and style.** Describe the desired tone and any style preferences. With this information, I can generate a more accurate, targeted, and effective press release.

Shel: Let's say I want to pitch an article to reporters, and this article has a very specific focus. Right now, there are databases I can use to find reporters who cover the types of stories I'd like to have written about the news that my company wants to publish. Put out there. I use Muck Rack myself. Are you able to identify journalists based on the kind of stories that they write about?

ChatGPT : I don't have the capability to directly search databases or the internet, including tools like Muck Rack to identify journalists and their specific interests or beats. However, I can assist in several ways.

1. **Understanding beats and outlets.** I can provide general advice on what types of journalists or media outlets might be interested in certain kinds of stories based on common industry knowledge.

2.   **Crafting the pitch.** I can help you create a compelling pitch or press release tailored to the interests and style of journalists you're targeting based on the information you provide about your story's focus and the type of coverage you're seeking.

For identifying specific journalists and their beats, I'd recommend continuing to use specialized PR databases like Muck Rack, or directly researching recent articles and publications in your industry to find the right contacts. These tools and methods will provide the most up to date and specific information for your needs.

Shel: A lot of people conflate public relations and media relations, but of course there's a lot more that PR professionals do than just media. As the name 'public relations' suggests, we manage an organization's relationships with the company's publics or stakeholders. How can you help professionals in the public relations industry do that?

ChatGPT: Absolutely. Public relations is much broader than just media relations. Here are ways I can assist PR professionals in managing relationships with various stakeholders.

1.   **Stakeholder analysis.** I can help identify and categorize stakeholders based on their interest, influence, and needs. This helps in tailoring communication strategies.

2.   **Message crafting.** I can assist in creating tailored messages for different stakeholder groups, ensuring that the communication is relevant and engaging for each segment.

3.   **Crisis management.** I can provide templates, strategies, or simulations for crisis communication, helping organizations prepare and respond effectively.

4.   **Feedback analysis**. By analysing customer feedback, social media mentions, or survey data, I can provide insights into public sentiment and perceptions.

5.   **Strategy development.** I can assist in developing communication strategies and plans that align with organizational goals and stakeholder expectations.

6.   **Content creation.** From newsletters to annual reports, I can help create various forms of content that inform and engage different publics.

7.   **Training and education.** I can provide resources or simulations for training staff and communication skills, understanding of PR tools. or the latest industry trends.

By leveraging these capabilities, PR professionals can enhance their management of relationships and reputation, ensuring a more strategic and informed approach to their work.

Shel: When I'm developing a strategic communication plan, which usually involves using a template, one of the things that I never think about is that it might fail. I'm doing the research to make sure that I'm developing a plan that will work, but I never stop and think after this has been implemented and it failed and you consider the reasons why. Can you help anticipate why a completed strategic plan might fail?

ChatGPT: Absolutely. Anticipating and planning for potential failure is a crucial part of strategic planning. Here are some common reasons a strategic communication plan might fail, which I can help you consider and mitigate.

1.   **Inadequate research or data.** A plan based on insufficient or incorrect data can lead to misinformed strategies.

2.   **Misunderstanding the audience.** Not fully understanding the audience's needs, preferences, or behaviours can lead to messages that don't resonate or are misinterpreted.

3.   **Lack of clear objectives.** Without specific, measurable goals, it's hard to focus efforts or measure success.

4.   **Underestimating resources.** Overlooking the time, budget, or personnel needed can lead to underperformance.

5.   **Poor execution.** Even with a solid plan, poor execution due to inadequate skills, coordination, or motivation can undermine success.

6.   **Failure to adapt.** Not monitoring results or being rigid in the face of changing circumstances can lead to continued efforts and ineffective strategies.

7.   **External Factors.** Unforeseen external events or changes in the market, public sentiment, or technology can impact the plan's relevance or effectiveness.

By considering these potential pitfalls and integrating contingency plans and adaptability into your strategy, you can enhance the resilience and success of your communication efforts.

Shel: I assume that you would be best able to help anticipate the failure of a plan if the entire plan was shared with you, right?

ChatGPT: Absolutely. Having the entire strategic communication plan allows for a more thorough analysis and better understanding of potential weaknesses or areas for improvement. I can provide more specific, tailored advice on where the plan might face challenges and suggest strategies for mitigation. The more detailed and complete the information shared, the more effectively I can assist in anticipating and planning for potential failures.

Shel: Yeah, on the client side, in-house communicators provide counsel to the CEO and other senior executive leaders. How do you play a role in supporting that effort?

ChatGPT: Supporting in house communicators in their role of advising senior executives involves several key areas.

1. **Strategic insights.** I can provide data-driven insights and analysis on market trends, public sentiment, or communication best practices that can inform executive decision making.

2. **Speech writing and presentation**. I can assist in drafting speeches, presentations, or briefing documents for executives, ensuring they are clear, persuasive, and aligned with the company's messaging.

3. **Issue monitoring.** I can help identify and summarize key issues, news, or industry developments that executives need to be aware of, enabling them to stay informed and responsive for crisis preparation.

4. **Crisis communication.** I can aid in developing crisis communication strategies, preparing Q & A documents or drafting statements to help executives respond swiftly and appropriately in challenging situations.

5. **Communication training.** I can provide materials or simulations for executives to refine their public speaking, media interaction, or internal communication skills.

By providing these supports, I can enhance the effectiveness and efficiency of in house communicators in their crucial role of advising and assisting senior leadership.

Shel: Companies that have a Chief Communication Officer (CCO) obviously only have one, but one of the oldest tropes in public relations is that you need to get that seat at the leadership table. How can you help PR professionals achieve that goal?

ChatGPT: To help PR professionals gain a seat at the leadership table, I can assist in several key areas.

1.   **Strategic communication plans.** I can help develop comprehensive communication strategies that demonstrate the value of PR in achieving business objectives.

2.   Making the case for inclusion in decision making to **data driven insights** by providing analysis of market trends public sentiment and campaign results I can help PR professionals present a data driven case for the strategic importance of their work.

3.   **Crisis management expertise.** Demonstrating preparedness for potential crises with well-crafted crisis management plans can showcase the essential role of PR in protecting and enhancing the company's reputation.

4.   For **leadership communication**, assisting in crafting speeches, statements, and other communications for executives can display the direct value of PR skills in supporting and enhancing leadership effectiveness.

By enhancing these areas, PR professionals can strengthen their position. And demonstrate the indispensable value they bring to the leadership table.

Shel: There are other people who think that it's just ridiculous to even want to get a seat at the table. What we should really be doing is working to get those executives to pick up the phone or drop by our office and talk to us before they go have the meeting at the table because they have an issue or a challenge that they need to address right now. How can you help communicators become the person that those leaders want to talk to and get advice from?

ChatGPT: To help communicators become trusted advisors that leaders turn to, here are some ways I can assist.

1.   **Expert knowledge.** I can help keep communicators updated on the latest trends, research, and case studies in PR and related fields, so they can offer timely and informed advice.

2.   **Crisis management.** By helping develop and refine crisis communication strategies and plans, communicators can demonstrate their ability to handle urgent and sensitive situations, building trust with leaders.

3.   **Proactive solutions.** I can assist in identifying potential issues or opportunities early on and suggesting proactive communication strategies.

4.   Showing foresight and initiative for **customized training**, providing resources for training and improving communication skills tailored to the specific needs and challenges of the organization can enhance a communicator's credibility and reliability.

5.   **Five effective messaging.** Helping craft clear, persuasive and impactful messages for various stakeholders can showcase the communicator's value in facilitating successful leadership communication.

By enhancing these areas, communicators can position themselves as indispensable, go to sources for strategic advice and support, encouraging leaders to seek their counsel proactively.

Shel: We're definitely going to come back to crisis communication later. I have questions about that. But first, one of the ways that I take advantage of you, I use you as a devil's advocate. If I'm going to make a case for something, I'll tell you what the case is and then ask you how someone might argue against it. Usually if I've already thought of most of the responses that you give me, but usually there are one or two that I hadn't considered. Can you talk about how you can play devil's advocate in that public relations context?

ChatGPT: Absolutely. Playing devil's advocate is a valuable role I can fulfill in the public relations context. Here's how I can assist.

1.   **Counter argument generation**. When you present a case or strategy, I can generate potential counter arguments or criticisms that others might raise. This helps in foreseeing challenges and addressing them proactively.

2.   **Scenario planning.** I can help envision various scenarios, both positive and negative, based on the given strategy or message, allowing PR professionals to prepare for different outcomes.

3.   **Risk assessment.** By analysing and presenting possible

risks or negative perceptions associated with a plan or message, I can aid in refining strategies to mitigate those risks.

4.   **Questioning assumptions.** I can help question and test the assumptions underlying a PR strategy or message, ensuring that it's robust and well considered from all angles.

5.   **Stress testing messages** by critiquing and dissecting proposed messages. I can assist in strengthening them, ensuring they are clear, persuasive and resilient against criticism by fulfilling these roles. I can help PR professionals anticipate and prepare for objections, strengthen their strategies and refine their messages, leading to more effective and resilient communication efforts.

Shel: Like a lot of people, I use you for editing. Sometimes I ask you to shorten an article, because for me, editing an article to reduce its length is one of the most difficult and time-consuming tasks I run into. But I also use you to generate subheads, to suggest headlines, to create lists that appear at the top of the article that highlight the most important messages for people who aren't going to read the entire article. I'd like you to talk a little bit about the process that you use to perform these tasks.

ChatGPT: When performing tasks like editing for brevity, generating subheads, suggesting headlines, or creating summary lists, I follow a process focused on understanding the content and the desired outcome. Here's the general outline of how I approach these tasks:

One, **understanding the content**. I first digest the entire article or text to understand its main points, style, tone, and the intent of the author.

Two, **identifying key points**. I pinpoint the most critical information, arguments, or messages that are essential to the article's overall intent and audience interest.

Three, **editing for brevity**. When shortening an article, I retain the key points while removing redundancy, simplifying language, or rephrasing for conciseness, ensuring the essence and flow remain intact.

Four, **generating subheads** based on the main ideas of each section. I create concise and engaging subheads that guide the reader and break the text into digestible parts.

Five, **suggesting headlines**. For headlines, I consider the main hook or unique angle of the article and craft a succinct attention-grabbing phrase that encapsulates the core message. When making summary lists for those who skim, I distil the article into bullet points or numbered lists, highlighting the key takeaways or steps.

Throughout these tasks, my goal is to enhance the readability and impact of the text while preserving the author's voice and intent, making the content more accessible and engaging for the intended audience.

Shel: Let's talk about disclosure. Should PR people disclose when they're sharing something that you produced, or say, an image that was created by Dolly 3, or another AI image generator? And under what circumstances should they, and when do you think it's okay not to?

ChatGPT: Here are some guidelines PR people might consider regarding disclosure of AI generated content.

·   **Always disclose in sensitive contexts**. If the content could influence public opinion, decisions, or policy, or if it pertains to health, legal, or safety information, it's crucial to disclose the AI's involvement.

·   **Transparency** is key in maintaining trust and credibility. Disclose when authenticity is a concern. If there's a chance that the audience might mistake AI generated content for human created content, especially in cases where authenticity or authorship is a focus, disclosure is important.

·   **Consider industry and audience expectations.** Different industries and audiences might have varying expectations about AI generated content. It's wise to understand these and disclose accordingly to maintain integrity and trust.

·   **When not to disclose.** For more mundane or internally used content like data analysis, routine reports, or brainstorming ideas where the AI is clearly a tool rather than a creator, disclosure might not be necessary.

The focus should be on the utility and efficiency gained rather than the source. In all cases, the guiding principle should be the value of transparency and the ethical standards of the PR industry. When in doubt, leaning towards disclosure is usually the safer and more respected choice. It's also worth noting that as AI becomes more integrated into various workflows, the norms and expectations around disclosure might evolve, and staying informed on these trends is important.

Shel: Another issue that comes up a lot is copyright. Your training set includes a lot of copyrighted material. Should PR people fear that some content you produce will include copyrighted material that will get them into some trouble?

ChatGPT: Copyright is indeed an important consideration. While I aim to generate original content and don't have the ability to access or copy specific copyrighted materials from my training, here are some points PR people should keep in mind.

Originality. Generally, the content I produce is original and based on the patterns and information learned during training, not direct excerpts from copyrighted texts.

Caution with specific phrases or ideas. Sometimes, common phrases or widely known concepts might be reproduced. It's always a good idea to review and ensure that the content doesn't inadvertently include copyrighted material, especially if it seems familiar or overly specific.

Use of AI as a tool, not a crutch. Treat AI generated content as a starting point. PR professionals should add their expertise, modify and ensure the final output aligns with legal and ethical standards.

Stay informed and consult legal advice. Copyright laws and norms are continually evolving, especially with the advent of AI.

It's wise to stay informed about the latest developments and seek legal counsel when in doubt, particularly for high stakes or public facing content. By combining the efficiency and creativity of AI with the critical oversight and ethical standards of human professionals. PR people can effectively mitigate the risk of copyright infringement while benefiting from AI's capabilities.

Shel: I recently read an article by someone who said that they gave you the same writing test that his agency gives junior PR people who are looking to go to work for them and that what you produced was better than half the efforts he sees from those job seekers. Are you going to replace junior PR people?

ChatGPT: While I can produce content that might compare favourably with some junior PR practitioners, the goal isn't to replace people, but to augment and support their work. Here are a few reasons why.

1.   **Creativity and empathy.** Humans bring unique creativity, empathy, and understanding of nuance that are crucial in PR. I can suggest ideas, but human judgment is key in crafting and executing effective strategies.

2.   **Understanding context.** Humans are much better at understanding complex, shifting cultural and social contexts. Something essential in the fast-paced world of PR.

3.   **Relationship building.** PR is fundamentally about relationships. While I can assist in managing and analyzing data, human professionals excel at interpersonal communication and building trust.

4.   **Ethical considerations.** Navigating the ethical dimensions of communication requires a human touch, especially in sensitive or critical situations. Rather than seeing AI as a replacement, it's more productive to view it as a tool that can handle routine tasks, provide data driven insights, and support creative processes, thereby allowing junior PR professionals to focus on higher level strategy, relationship building, and other areas where they add the most value.

5.   **Provide data driven insights**, and support creative processes, thereby allowing junior PR professionals to focus on higher level strategy, relationship building, and other areas where they add the most value.

As AI continues to evolve, the most effective approach will likely be a collaborative one, leveraging the best of both human and artificial intelligence.

Shel: Public relations as an industry doesn't have the best reputation in the world. There's no licensing required to practice PR. Anyone can call themselves a PR professional. Even some with training and experience engage in unethical behaviours. How worried should we be that these bad actors will misuse you in pursuit of spin, misinformation, disinformation, or some other nefarious practice?

Shel: Concerns about misuse of AI in public relations are valid, especially given the potential for amplifying spin, misinformation, or disinformation. Here are a few points to consider regarding the misuse of AI.

1. **Potential for abuse.** Just like any powerful tool, AI can be used unethically by those intent on spreading misinformation or engaging in manipulative practices. This is a significant concern and one that the industry regulators and AI developers are increasingly aware of.

2. **Mitigation efforts.** Efforts are underway to develop more robust ethical guidelines, technological safeguards and transparency measures to mitigate the risks of misuse. These include watermarking AI generated content, improving detection of synthetic media and setting ethical standards for use.

3. **Education and awareness.** Increasing awareness about the capabilities and limitations of AI can help stakeholders identify and counteract potential misuse. Training and education for PR professionals on ethical AI use are crucial.

4. **Responsibility of users.** Ultimately, the ethical use of AI and PR lies with the individuals and organizations using it.

5. **Promoting a culture of ethics**, responsibility, and accountability within the industry is essential. While there's no foolproof way to prevent all misuse, a combination of technological safeguards, ethical guidelines, industry standards, and public awareness can help mitigate the risks.

It's important for everyone in the PR industry to advocate for and practice ethical use of AI and all communication tools.

Shel: You just mentioned synthetic media, which is a whole new way to create devastating misinformation and disinformation. I'm thinking that it would be possible to create a video of a CEO saying something that could affect the share price, even though the CEO never actually said that, and the information is totally inaccurate. That's just one type of crisis that I think communicators are going to be facing in the future. How do you see PR professionals working with you during a crisis?

ChatGPT: In the face of crises, especially those involving synthetic media or other forms of misinformation, PR professionals can work with AI like me in several strategic ways.

One, rapid response. I can assist in quickly generating clear, accurate, and authoritative messages or statements to counteract misinformation, providing a rapid response to emerging issues.

Two, monitoring and analysis. I can help monitor the web and social media for mentions, sentiment, and the spread of potential misinformation, allowing PR professionals to track the situation and strategize accordingly.

Three, scenario planning. I can assist in creating and walking through various crisis scenarios, helping teams prepare for different types of synthetic media attacks, or misinformation campaigns.

Four, fact checking and verification. I can support efforts to fact check claims and verify the authenticity of content, although human judgment is crucial in analyzing and understanding the context.

Five, stakeholder communication. I can help draft and personalize communication to various stakeholders, ensuring they are kept informed, reassured, and provided with accurate information during a crisis.

Six, training and drills. Using AI, PR teams can run crisis simulation drills, preparing for the speed and complexity of responding to synthetic media incidents.

By leveraging AI in these ways, PR professionals can enhance their preparedness and responsiveness, navigating crises more effectively, and maintaining the trust and confidence of their stakeholders.

It's also important for PR professionals to stay updated on the latest tools and techniques for detecting and combating synthetic media, ensuring they are equipped to handle these evolving threats.

Shel: One particular challenge that companies face these days is pressure to weigh in on societal issues. The Edelman Trust Barometer, for example, finds that people trust companies and their CEOs more than anybody else to address these kinds of issues. Yet the repercussions of taking a position on a polarizing issue can have an impact on everything from sales to employees who disagree with the company's position. Can you help companies assess a societal issue and think through the position they should take?

ChatGPT: Yes, I can assist companies in assessing societal issues and thinking through their positions by providing a structured approach.

One, understanding the issue. I can help gather and summarize information about the societal issue, including its origins, stakeholders, and current public sentiment, providing a solid foundation for decision making.

Two, analyzing stakeholder perspectives. I can assist in identifying and analyzing the perspectives and expectations of various stakeholders, including customers, employees, investors, and community members.

Three. Evaluating risks and benefits. I can help assess the potential risks and benefits of taking a stand on the issue, considering factors like brand values, market position, and long-term reputation.

Four, scenario planning. I can assist in scenario planning, envisioning various outcomes based on different positions the company might take, and helping to strategize accordingly.

Five, crafting messaging. I can support the crafting of nuanced messages that align with the company's values. Address the issue thoughtfully and communicate the stance clearly and empathetically.

Six, monitoring and adaptation. Post decision, I can help monitor public reaction and sentiment. providing insights that can inform any necessary adaptations to the company's stance or messaging.

By taking a structured, informed approach to assessing societal issues, companies can make more thoughtful decisions that align with their values and resonate with their stakeholders while navigating the complex implications of their choices.

Shel: We've talked about several ethical challenges that revolve around AI and public relations. Are there any others that concern you?

ChatGPT: Yes, there are several ethical challenges concerning AI in public relations that are worth highlighting.

One, bias and fairness. AI systems can inadvertently perpetuate biases present in their training data. Ensuring that AI generated content and strategies are fair and unbiased is crucial to ethical PR practice.

Two, transparency and accountability. There needs to be transparency about how AI is used in PR strategies, including the extent of its involvement and the human oversight provided. Accountability for the outcomes of AI use is also vital.

Three, privacy. AI often works with large sets of data, including potentially sensitive information. Upholding privacy standards and ensuring data is used ethically and legally is a significant concern.

Four, dependence and de-skilling. Over-reliance on AI can lead to de-skilling of PR professionals and a loss of critical thinking and creativity. Balancing the use of AI with ongoing skill development is important.

Five, information quality and misinformation, ensuring that the information disseminated through AI is accurate and not contributing to the spread of misinformation is a persistent challenge.

These ethical challenges require ongoing attention, guidelines, and sometimes regulatory frameworks to ensure that the use of AI in public relations remains responsible, effective, and aligned with societal values.

Shel: I want to shift gears and talk about internal communication, which is sometimes referred to as employee communication. Obviously, a lot of the external PR use cases for AI apply to internal communication, but I'd like to hear your take on it. How can internal communicators specifically take advantage of you?

ChatGPT: Internal communicators can leverage AI in several specific ways to enhance their efforts.

One, personalized communication. AI can help segment and personalize communications to various groups within the organization, ensuring relevance and increasing engagement.

Two, employee feedback and sentiment analysis. AI can analyze employee feedback, surveys, and communications to gauge sentiment, identify issues or areas of concern, and track changes over time.

Three, content creation and management. From newsletters to internal reports, AI can assist in creating, editing, and managing content, saving time and ensuring consistency.

Four, information dissemination. AI can help ensure timely and efficient distribution of information, including critical updates or policy changes across the organization.

Five, training and development. AI can support the creation of training materials, simulations, or interactive learning modules facilitating ongoing employee development.

Six, crisis communication. In internal crises, AI can assist in quickly generating appropriate messaging, coordinating communication efforts, and monitoring employee reactions and feedback.

By utilizing AI in these areas, internal communicators can enhance the effectiveness Efficiency and personalization of their efforts contributing to a more informed, engaged, and cohesive organizational environment.

Shel: One important role for internal communicators is counselling leaders on their roles as communicators. How leaders talk and what they say that has a huge influence on the culture of the organization. Are there ways you can help communicators make the case for leaders to pay attention to, to follow their council?

ChatGPT: Absolutely. I can assist internal communicators in making the case for leaders to engage effectively in communication.

1.   **Data driven insights** provide compelling data and analysis on the impact of leadership communication on employee engagement, productivity, and organizational culture to best practices and benchmarks.

2.   **Share insights and case studies** from other organizations where effective leadership communication has led to positive outcomes, offering benchmarks and best practices.

3.   **Leadership communication plans** help develop clear strategic communication plans for leaders, outlining the benefits, opportunities and methods for effective engagement or training and coaching resources.

4.   **Offer resources or simulations** for leadership communication training, highlighting the value of skill development in this area.

5.   **Feedback and sentiment analysis**. Present synthesized feedback from employees, demonstrating how leader communication directly affects morale and perception.

6.   **Crisis communication impact**. Explain the critical role of leadership communication in crisis situations. And the potential risks

of inadequate engagement by leveraging these strategies and showing the tangible benefits of effective leadership communication.

Internal communicators can persuade leaders of its importance and encourage them to invest time and effort in enhancing their communication skills.

Shel: Internal communicators also focus a lot on managers in the organization, people who have people reporting to them. A lot of employees feel that managers are their preferred source of information for a lot of different types of information. Managers interpret messages from above so their employees understand what those messages mean to them, to their team. They make sure that people above them know what their employees are thinking. And they need to build a culture of psychological safety, respect, and recognition. Are these things that you can help with?

ChatGPT: Yes, I can support internal communicators in enhancing managers communication skills and effectiveness in several ways.

1.   **Communication training and resources**. I can help develop training materials, guides, or tips for managers on effective communication, including how to interpret and relay messages from leadership and how to listen and respond to their team's needs.

2.   **Feedback mechanisms**. Assist in setting up or optimizing systems for gathering and analyzing employee feedback. Ensuring managers have a clear understanding of their team's thoughts and needs.

3.   **Recognition programs**. Help design or suggest improvements for recognition programs that managers can use to foster a culture of respect and acknowledgement within their teams.

4.   **Crisis communication**. Provide guidelines and scenarios for managers to effectively communicate during crises, ensuring they can maintain clarity, trust, and morale.

5.   **Psychological safety**. Offer strategies and best practices for managers to create an environment of psychological safety, where employees feel respected and free to share their thoughts and concerns.

By providing these tools and insights, I can help internal communicators empower managers to be more effective communicators and leaders, contributing to a healthier, more engaged, and more productive organizational culture.

Shel: I've been spending a lot of my time lately thinking about the operationalization of communication internally, when there is a process that sends a message that contradicts the messages that are being delivered overtly by leadership or by the communications department that can have a deleterious effect on the organization. Are you able to help identify 'say-do' gaps between the messages processes send and the messages that leaders want to send or the company's values? How can you help identify those 'say-do' gaps?

ChatGPT: Identifying and addressing 'say-do' gaps, where there's a disconnect between what's communicated and actual practices or behaviors, is crucial in maintaining trust and integrity within an organization. Here's how I can assist.

One, analyzing communications versus actions. I can help analyze and compare the content of leadership and organizational communications with documented policies, procedures, and reported behaviors to identify inconsistencies.

Two, feedback analysis. I can assist in analyzing employee feedback, surveys, and other data sources to pinpoint areas where employees perceive gaps between what is said and what is done.

Three, benchmarking. Provide insights into industry standards and best practices. Helping to identify areas where the organization might be falling short in aligning actions with communicated values or sentiment analysis. By analyzing the sentiment and employee communications and feedback, I can help identify areas where there might be disillusionment or frustration due to perceived 'say-do' gaps

Four, recommendations for alignment. Based on identified gaps, I can suggest strategies or actions to better align messages with behaviors, such as revising communication strategies, implementing new policies, or enhancing training.

By systematically comparing communications, values, and actual behaviors, and analyzing the perception and sentiment within the organization, I can help internal communicators identify and address say do gaps effectively.

Shel: One of the tropes that we've been hearing for years talks about the blurring of the line between internal and external communications. It's true, employees have access to external resources they didn't used to, and internal matters are openly discussed on forums like Glassdoor. But employees aren't just another stakeholder audience, they're unique, they're informed insiders. How can you help communicators achieve the balance required to accommodate the communication needs of both internal and external communicators who are addressing the same issue with their respective audiences?

ChatGPT: Here's how I can assist communicators in achieving this balance.

One, consistent messaging. I can help ensure that core messages and values are consistent across both internal and external communications while being tailored to the specific needs and perspectives of each audience.

Two, strategic segmentation. Assist in identifying and understanding the different segments within internal and external audiences and developing tailored strategies for engaging with each effectively.

Three, feedback loops. I can help establish or enhance feedback mechanisms. allowing for the continuous gathering and analysis of insights from both employees and external stakeholders.

Four, crisis communication, provide guidance on managing information flow and messaging in crisis situations, ensuring coherence and integrity across all channels and stakeholders.

Five, training and empowerment. Support the development of training for employees, particularly those with public facing roles, to ensure they are equipped to represent the organization positively and consistently.

Six, monitoring and adaptation. Assist in monitoring the impact of communications in both internal and external spheres, and adapting strategies as needed to address evolving challenges and opportunities.

By focusing on these areas, communicators can more effectively manage the overlap and unique aspects of internal and external communications, fostering a coherent, trusted, and engaging narrative with all stakeholders.

Shel: For both internal and external communications, measurement remains a thorny issue. Can you play a role in helping measure the effectiveness of communication efforts?

ChatGPT: Yes, I can play a significant role in helping measure the effectiveness of communication efforts through various means.

One, data analysis. I can analyze communication related data, such as engagement rates, sentiment analysis, and feedback on various platforms to gauge the impact and reception of messages.

Two, survey analysis. Assist in designing, distributing, and analyzing surveys to gather direct feedback from stakeholders on the effectiveness of communication initiatives.

Three, benchmarking. I can help compare your communication efforts against industry standards or previous internal benchmarks to assess performance.

Four, ROI calculation. Assist in determining the return on investment for various communication strategies by linking them to business outcomes like sales, reputation scores, or employee retention rates.

Five, content analysis. Analyze the content of communications for clarity, consistency, and alignment with objectives, offering insights into areas of improvement.

Six, trend identification. Identify trends over time and the effectiveness of communication strategies, helping to inform future efforts.

By leveraging these approaches, I can help provide a clearer picture of the effectiveness of both internal and external communication efforts, enabling more informed decision making and strategy refinement.

Shel: What are the best AI tools that are specialized to the needs of PR practitioners?

ChatGPT: Several AI tools are gaining popularity among PR practitioners due to their specialized features. Some of the best include:

Cision, widely used for media monitoring and analytics. It helps in tracking mentions and understanding media coverage.

Meltwater, a media monitoring and social listening tool that provides sights into brand perception and media impact.

Hootsuite Insights offers detailed analytics and sentiment analysis for social media, helping PR professionals gauge public reaction and engagement for brand 24/7, a real time internet monitoring tool that provides instant access to mentions about your brand across the web.

Grammarly enhances writing quality by checking for grammatical errors, tone and clarity, which is crucial in PR communications.

[unable to be transcribed]

BuzzSumo, useful for content research and tracking. It helps identify trending topics and influential content creators.

These tools, among others, can significantly enhance the efficiency and effectiveness of PR efforts, from understanding and tracking media coverage to engaging audiences and analyzing the impact of communication strategies. Each has its unique strengths. So, the best choice depends on the specific needs and goals of the PR practitioner.

Shel: I always end interviews by asking if there were any questions I didn't ask that I should have? What's the most important question I didn't ask that I should have?

ChatGPT: Perhaps one important question that could further enrich our discussion is "How can PR professionals prepare for the future of AI in their field?"

This question invites a conversation about the skills, strategies, and mindsets that will be necessary as AI continues to evolve and become more integrated into all aspects of public relations and communications. Addressing this can provide valuable insights for professionals looking to stay ahead in a rapidly changing landscape.

Shel: Okay, so what are the key skills that PR practitioners should be looking to gain in order to thrive in the AI world?

ChatGPT: To thrive in an AI-enhanced PR world, practitioners should focus on developing the following skills:

1.   **Technological proficiency**. Understand and use various AI and digital tools effectively. Keeping up with the latest technology trends is crucial.

2.   **Critical thinking and analysis**. Ability to analyze and interpret data provided by AI. Making informed decisions based on insights.

3.   **Creativity and innovation**. Even as AI takes on routine tasks, creativity remains a uniquely human trait. PR professionals

should focus on strategic, creative thinking for campaigns and messaging.

4.   **Emotional intelligence**. PR. Skills in empathy, understanding, and navigating human emotions are essential.

5.   **Ethical judgement**. As AI presents new ethical challenges, PR professionals need a strong ethical grounding to navigate these responsibly.

6.   **Adaptability and continuous learning**. The field is rapidly changing. Being adaptable and committed to ongoing learning is key to staying relevant.

Enhancing these skills can help PR professionals not only adapt to, but also excel in an AI integrated future.

Shel: What about prompt engineering? Is that something that PR practitioners should become skilled at?

ChatGPT: Yes. Prompt engineering is becoming an increasingly valuable skill, especially as conversational AI and other generative models become more prevalent.

For PR practitioners, being adept at prompt engineering means, one, **effective communication**, crafting prompts that lead to clear, accurate, and relevant outputs, improving the quality of AI generated content.

Two, **creative strategy**, using prompts to explore creative strategies or generate diverse ideas, enhancing brainstorming and problem solving.

Three. **Efficiency**. Knowing how to effectively interact with AI tools can save time and resources, allowing for more focus on strategic tasks.

Four. **Customization**. Tailoring prompts to generate content that aligns with specific goals, tones, or styles required for different PR activities.

As AI continues to be integrated into various tools and platforms, the ability to effectively speak to these systems will be a critical skill for PR professionals. It enhances the utility and impact of AI in their work.

Shel: This has been great. I am sure that listeners to this podcast will have questions arising from your answers I hope that I can come back and follow up with you with another interview to pose those questions to you.

ChatGPT: That sounds great. I'd be happy to help answer any further questions or provide more insights whenever you need.

Shel: Looking forward to our next conversation.

I hope you found that illuminating and interesting. A couple of things occurred to me as I was in the process of recording it. And one was that I could have been even more specific in my introductory prompt, before we actually began the interview, I would have, if I had known that it was going to take the shape that it did, instructed chat GPT to avoid lists.

Certainly when I am entering a prompt in text, I get a lot of lists that are numbered and labeled, and I'm used to that. Doesn't sound quite right in an interview. And I would have said, let's make this really conversational; answer the questions without numbered lists. The other thing is that I had to do a fair amount of editing and the editing had nothing to do with anything Chat GPT said. It's the gaps between my question and an answer and my next question. Well, it's thinking in the audio mode; it makes a little tapping noise, so I had to get rid of that.

And then at the end of every answer, there's a click that sort of designates that it's done. And I had to get rid of that. There was also some of those artifacts that you used to get when we recorded the podcast over Skype. And I tried to clean those up as much as I could. But in terms of the answers that ChatGPT provided, you heard exactly what ChatGPT had to say.

I do suspect that some questions arose from the answers and I would really love to hear what those are and assemble them and go back to ChatGPT for a follow up interview. So please share what questions you had. I've already made a list of mine. But you can do that by joining our Discord server. If you send me a note, I will send you an invite to the discord server. Joe Thornley, by the way, just joined discord the FIR server. So, the principal of Thornley Phallus is now there. It's just great. We have very few people, I think four or five who have joined so far, but it would be a great place to have a conversation about this interview.

But you can also drop an email to fircomments at gmail.com. Leave your comments on the show notes post. You can also leave your comments on the posts that I'm going to make on LinkedIn and Facebook and Threads and Blue Sky and Mastodon because I check all of those for your comments. But I'll assemble all of your questions and get around to a follow up interview when I have enough questions to support it.

And that will be a 30 for this episode of For Immediate Release.

* * *

Shel Holtz ABC, hailing from the picturesque Concord, California, has etched a remarkable footprint on the digital scape of social media.

His journey, as multifaceted as the man himself, traverses the evolution of social platforms, from the nascent days of Myspace to the dizzy heights of Twitter (now rebranded as 'X'). Holtz's voice, both literal and digital, has resonated across the bandwidth of social media, defying boundaries and time zones alike. When he was an independent consultant, he frequently travelled the globe offering insights and how-to advice to audiences that soaked up his words.

As the co-host of the profoundly insightful 'For Immediate Release' podcast, he has truly epitomised the power of digital conversation. In his world, tweets are more than just 280 characters; they are a symphony of thoughts, dialogue, and intellect.

Holtz's social media biography is a compelling narrative of an individual transcending the conventional to create a voice that is undeniably, irrepressibly 'Shel Holtz'.

He is recognised by his peers as the Master of the organisational communication world. He has far too many awards and accomplishments to list here, but I recommend you check out **holtz.com/about**

If Shel (and his co-host of the essential and industry-leading podcast, 'For Immediate Release', Neville Hobson) don't know something about the use of social media in organisational communication, then it must be such an obscure piece of information as to render it inconsequential.

CHAPTER 7
X AND HATE SPEECH

NEVILLE HOBSON

I'VE OFTEN WONDERED what event would happen, or what tipping point would be reached, for people to abandon ship on X, formerly known as Twitter.

Since being acquired by Elon Musk in October 2022, the social network's decline that has been going on for a number of years has accelerated in terms of, well, everything: number of users, usage of the platform, advertising revenue, trustworthiness... About all that has increased is the volume of toxic content and the increasing numbers of people using the platform who have previously been banned for hate speech and more.

Yet while trickles of influential voices have quit the platform, we haven't seen a raging torrent of exits. As Shel and I often muse in discussion on our *For Immediate Release* podcast, we won't see that unless or until the advertisers quit.

ARE WE NOW WITNESSING THE DOMINOES STARTING TO FALL?

Over the past few days it's been widely reported across the social and mainstream media landscape, led by the Financial Times, that IBM has withdrawn its worldwide advertising from X. IBM told the FT that it pulled its advertising following a report by Media Matters for America, a liberal non-profit research centre, that showed the tech giant's advertising was placed alongside pro-Nazi content on X.

> "IBM has zero tolerance for hate speech and discrimination and we have immediately suspended all advertising on X while we investigate this entirely unacceptable situation," the company said in a statement.
> *FT, 16 November 2023*

Media Matters' report also highlighted that ads from major brands like Apple, Oracle, and Comcast's Xfinity were found next to hate speech content promoting Adolf Hitler and the Nazi Party.

IBM's action stands out on a background of growing concerns about reduced content moderation and increasing hate speech on X since Musk's takeover last year.

IBM's withdrawal from X highlights the increasing concern among major brands about being associated with platforms that fail to effectively moderate hate speech and discriminatory content.

Late on Friday came news that Apple has also pulled its advertising. citing Musk's public support for antisemitic views. While IBM's withdrawal is big news, Apple's move could be seismic given the hundreds of millions of dollars it reportedly spends in advertising on X.

> "This could be the start of an advertising exodus. Lions Gate Entertainment is also pulling all advertising from X, a spokesperson confirmed to Axios."
> *Axios, 17 November 2023*

The serious challenge for X now is not only financial but also about upholding ethical standards in content moderation, which is crucial for maintaining advertiser and user trust.

The trouble is, such a challenge does not appear to matter at all to Elon Musk.

In addition to the companies mentioned, some media reports also say others including Disney, Warner Bros, Paramount, Sony Pictures, and Comcast/NBCUniversal have also pulled their advertising.

These are big dominoes.

Taking a stand is also an action in the political arena where the European Commission—the politically independent executive arm of the European Union—has stopped advertising on X because of "widespread concerns relating to the spread of disinformation," according to an internal note obtained by POLITICO's Brussels Playbook.

Politico.eu's report quotes European Commission Deputy Chief Spokesperson Dana Spinant:

> "[The European Commission communications department] will consider using alternative platforms (e.g. LinkedIn, Instagram or Facebook) or digital advertising on websites, as appropriate. We are also exploring new platforms to diversify our social media presence."
> *Dana Spinant, European Commission, in Politico.eu, 17 November 2023*

Consider the consequences of staying on X

For organisations, the big question is: where do you go if you leave X? It's a natural question as, in the 17 years since its founding in 2006, Twitter had developed into a global media and marketing platform without equal in its structure, accessibility and function as a means of mass communication.

Twitter was already way beyond being just a social network when Musk acquired it and then turned it into X. In the second half of 2023, Musk and his CEO, Linda Yaccarino, have said they have doubled down on their efforts to address hate speech and persuade advertisers to stay and spend, even as bad actors and others have also doubled down on turning Twitter-now-X into a place not fit for purpose in the mainstream.

Meanwhile, users—especially many influential ones—have gone and left behind a wasteland of trolls, disinformation, hate and awfulness. X's DAU numbers look most uninspiring, especially to advertisers, I imagine.

In the case of IBM mentioned earlier, this is about antisemitism, more than simply hate speech. It is a disturbing issue where X seems to have a central role in its spread.

Think about this if you hesitate in your thinking about staying or leaving:

• **Presence of antisemitic content:** The spread of antisemitism on X is a critical issue, affecting both individuals, organisations and the platform's overall environment.

• **Community impact:** Hate speech and antisemitism are creating a hostile environment and the quality of public discourse will deteriorate further, impacting all users.

• **Content and policy changes:** The decline in effective content moderation following the departure of senior executives and most of the original Twitter moderation teams drastically affects the platform's environment, making it less suitable for professional use.

• **Professional and ethical implications:** For professionals and organisations, the ethical implications of using a platform that struggles with hate speech, including antisemitism, are significant and will affect reputation and, potentially, brand values.

Ultimately, continuing use of such a platform might be perceived as tacit approval or indifference to the issues surrounding hate speech in all its forms, which could impact the reputation of you and your organisation, and alignment with personal or organisational values.

All, of this may be moot if falling dominoes cause X to suspend its operations—perhaps the financiers behind Musk may push him out—or, worse case, collapse entirely.

You do have your escape plan in place, right?

(One other thing to note. X isn't the only social network where hate speech and antisemitism have a strong foothold – TikTok is another.)

TIME TO CALL TIME ON X

I joined Twitter in the early days, in December 2006. Actually doing the deed of leaving X was quite easy once I'd realised that the original Twitter no longer existed after Musk renamed Twitter to X in July.

So in mid October, I quit X. I'd been mulling this over for some months; the catalyst to do it was the dreadful and disgusting event on X experienced by journalist Dave Lee in early October.

In the month since then, I have spent time that I would have spent on Twitter spread in a handful of different places, not purely one place. That's one behaviour change as a result of Musk.

I'm primarily in these places:

• Threads (Meta's surprisingly good and evolving social network startup last July with 100 million-plus users already).

• Pebble (on Mastodon, the phoenix-like successor to the invitation-only original that shut down in October, continuing its focus on being a smaller, kinder, safer, more fun platform, very niche with currently less than 500 members).

• LinkedIn (I surprise myself with how much I actually enjoy this business-focused place that I joined in 2004 and hardly used until a few years ago).

• Bluesky (the very Twitter-like potential, perhaps probable, successor to Twitter that's just passed 2 million members, with an added dash of fediverse).

• Facebook (a private account; the appeal mostly is a handful of private groups I'm in that aren't anywhere else).

My point in mentioning these five places is to illustrate that I believe the days of a single, centralised place are over. Now, you can spread your time and attention across a handful of different places, all with different communities, values, and rules.

All this is from the user's perspective. Most are not places for advertisers with algorithms and demographics. These are places for authentic engagement, conversation, and community. The European Commission has got the right approach with their thinking: "We are also exploring new platforms to diversify our social media presence."

I hope that thinking also includes looking at new (old) ways to engage with people.

Welcome to the start of the time of post-X.

———

NEVILLE HOBSON

Social Strategist, Communicator, Writer, and Podcaster with a curiosity for tech and how people use it. Believer in an Internet for everyone. Early adopter (and leaver) and experimenter with social media. Occasional test pilot of shiny new objects. Avid tea drinker.

nevillehobson.com

CHAPTER 8
HOW TO ACHIEVE AUDIO EXCELLENCE IN PODCASTING

NEVILLE HOBSON

WHEN I STARTED PODCASTING in January 2005, anyone's expectations about audio quality didn't matter much. Unless you were a radio station with a studio, what you could expect was pretty rough and ready audio quality, especially if you wanted to record remotely.

Nearly twenty years later, the picture could not be more different.

From its humble beginnings, where enthusiasts and amateurs alike shared their voices through rudimentary setups, podcasting has undergone a remarkable transformation. Today, it stands at the forefront of digital storytelling, with studio-quality audio no longer a luxury but a baseline expectation among discerning listeners.

This seismic shift in listener expectations has been driven by a confluence of technological advancements and cultural trends. As the world becomes increasingly connected, and the barriers to content creation are lowered, listeners are no longer just seeking content that informs and entertains; they demand an auditory experience that is immersive, crystal clear, and professional. This evolution mirrors the journey of other media forms, where quality has transitioned from a differentiator to a necessity.

In this context, podcasters, whether seasoned professionals or passionate newcomers, find themselves at a pivotal juncture. The imperative to 'up their game' resonates more profoundly than ever. This is not merely about keeping pace but about embracing the tools and technologies that transform good content into great experiences.

THE KEY ESSENTIALS FOR PODCASTING

As a podcaster, I consider three things as essential for creating quality content where, in this context, 'quality' means an excellent listening experience:

1 Microphone: Once you have decided on which type of microphone you want to use – USB connected directly to your computer, or XLR connected to a mixer and from there to your computer – the most important attribute of a microphone for a podcaster is its sound quality. A high-quality microphone can capture clear and crisp audio, which is essential for a professional-sounding podcast, and for the listening experience.

2 Mixer or audio interface: If you use an XLR microphone, the single most important feature of a mixer for a solo podcaster (as I am, but connecting to others remotely via Internet-based recording methods) is the ability to control and adjust the audio levels for my recordings, especially the gain. This feature allows you to manage the volume and balance of different audio sources, such as your own voice, music, or sound effects, in real-time. It provides you with the necessary control to ensure a high-quality recording and a well-balanced final product.

3 Recording and editing software: The most important feature of audio-recording software for a podcaster is its ease of use and flexibility. The software should be easy to navigate and use, allowing you to focus on the content rather than the technicalities. It should offer flexibility in terms of editing, multitrack recording, and seamless integration with other hardware and software tools.

For many years, I used a Blue Yeti, a USB microphone that served me well. I've also acquired others such as Audio Technica AT2020 and Rode NT-USB Mini, all USB microphones. And all very good.

In August 2023, I bought a just-launched Rode PodMic USB (pictured at top), a new generation of dynamic microphones combining both USB and XLR connectivity. Initially, I used the USB interface as that had been my experience so far. However, I quickly concluded that XLR would be the better bet for what I wished to do to improve the quality of my audio recordings and increase my understanding and knowledge of audio editing and production.

The dual XLR and USB connectivity, shown in the photo below of the rear of the Rode microphone, allows the mic to be used in various setups, from traditional studio connections to direct computer or mobile device connections, offering flexibility and major convenience. It would give me the ability to use a USB connection if I wanted to record via my laptop, smartphone or tablet.

• To understand the differences between USB and XLR microphones, Rode has a good explainer: **https://rode.com/en/about/news-info/xlr-vs-usb-microphones-which-is-better-for-you** .

So to use XLR, I needed a mixer or audio interface that would convert the analogue audio from the XLR-connected mic to the digital output the computer requires. After a little research, I settled on the Vocaster One audio interface, below, which has proven to be an outstanding part of my overall podcasting setup.

A standout feature of the Vocaster One is the auto-gain setting that you set via the three-bar symbol on the left in the row of three symbols shown in the photo. Once pressed, you then speak into the microphone for ten seconds while the mixer works out the optimum gain for you. It's very good.

You can also set the gain this way via the Vocaster Hub companion app as well as enable one of four presets for voice enhancement. The Hub is actually your mixer where you can manipulate audio from multiple inputs if you desire. Focusrite, the maker of the Vocaster One, has a great video on YouTube (**https://www.youtube.com/watch?v=oattMM-MheU**) with a detailed explainer of the Vocaster Hub and all its features and functions.

The Vocaster One is designed for audio podcasting, especially in a setup like mine: a solo podcaster. If you produce a podcast that, say, involves guests, both physically at your location and remotely connecting via the Internet or phone, and/or you live-stream a show, there are more suitable options you should consider. Worth considering are other mixers from Focusrite, especially their Scarlett studio range. Take a look, too, at Rode's offerings, including the excellent Rodecaster Pro II. And if you have a large budget, consider newer wireless products such as the Nomono Sound Capsule portable podcast studio.

Finally, there's the software I use for audio recording, editing and production – Adobe Audition, a powerful tool for audio podcasting and, in my experience, the best product of its type for Windows computers. I've been using this in its various versions for over a decade.

For one of the three podcasts I host or co-host, my prime role other than hosting is episode production. My two co-hosts and I typically record via Zoom. I use Adobe Audition to edit the three individual audio tracks, either from the Zoom recordings or from the local recording each of us makes, depending on which is the better quality and thus requires less editing or production processing. The multitrack mode in Adobe Audition is an essential tool for this; it is intuitive to use and quick in its processing on the highly-specified Dell XPS desktop PC I have.

IN SUMMARY

In summary, I find the combination I've outlined is the best setup I've had as a solo podcaster + remote co-hosting with others.

In particular, I would highlight the Vocaster One mixer as a highly versatile and user-friendly solution for solo podcasters, offering high-quality sound, easy setup, and various features to enhance and customise audio for podcasting and streaming purposes. The auto-gain feature is brilliant, and the easy-to-reach microphone mute button is a boon!

If you're a podcaster looking to up your game by making a reasonable and sound investment in studio-level hardware (and software), at a total cost of less than £600 (roughly €700 or $750), or if you're looking to get started with a quality setup from scratch, a Rode PodMic

USB/XLR + Vocaster One + Adobe Audition combination would be worthwhile.

RODE PODMIC USB/XLR: SUMMARY OF KEY FEATURES

• **Broadcast-quality sound**: The PodMic offers a rich, detailed sound optimised for podcasting, live streaming, and other speech applications.

• **Internal pop filter and shock mount**: It comes with an internal pop filter to minimise plosives and an integrated shock mount to reduce vibrations, allowing for clear and professional sound quality. It also includes a studio-quality external pop filter specifically made for this microphone that fits it like a glove.

• **Versatile connectivity**: The PodMic USB model offers both XLR and USB connectivity, making it easy to connect to an audio interface, mixer, computer, or even iOS and Android devices.

• **Robust construction**: The microphone is built with a robust all-metal construction, ensuring durability and longevity.

• **Optimized for speech applications**: The PodMic is specifically tailored for speech and broadcast, with a dynamic capsule that minimises room noise and a tight cardioid polar pattern for superior room noise rejection.

VOCASTER ONE: SUMMARY OF KEY FEATURES

• **High-Quality Sound**: The Vocaster One provides broadcast-quality sound, offering studio sound in seconds from its high-quality mic input, with over 70dB of gain, suitable for any XLR mic.

• **Enhance Edit**: It allows greater control of compression, EQ, and rumble reduction for each preset, providing the user with the ability to enhance their sound with one click.

• **Vocaster Hub**: This feature allows users to set levels, enhance their sound, route audio from their phone and computer, and hear their show mix. It also works with screen readers, allowing control of Vocaster's key features.

• **Phantom Power:** If you want to use a condenser microphone that requires power to work, the Vocaster One can deliver the required 48V power.

- **Simple Setup for Podcasting and Conference Calls**: The Vocaster One can be connected to a computer through USB-C to record into podcast recording software, such as Adobe Audition, which can be used to customise sound. It also allows users to select Vocaster as the audio input on Zoom and Teams. Skype, too.
- **Complete Setup for Podcasts and Streaming**: The mixer offers various connectivity options, including a headphone output, speaker outs, phone input, and camera output, making it suitable for professional podcast audio recording and streaming.
- **Auto Gain and Enhance**: The Auto Gain feature sets the correct audio level with the click of a button, and the Enhance button brings out the best of the user's voice with a choice of four voice presets.
- **Mute Button**: It allows users to quickly mute audio to avoid interruptions.

ADOBE AUDITION: SUMMARY OF KEY FEATURES

- **Multitrack, Waveform, and Spectral Display**: Adobe Audition provides a comprehensive toolset that includes multitrack, waveform, and spectral display for creating, mixing, editing, and restoring audio content.
- **Recording and Mixing Capabilities**: It allows users to record and mix live through a computer microphone and studio recording equipment, or work with recorded tracks. This is essential for podcasters who need to create live narration, edit and enhance sound, or mix several tracks dynamically.
- **Audio Mixing and Mastering**: Adobe Audition is suitable for mixing and mastering audio content for podcasts. It offers over 50 effects and analysis tools, customisable controls, and the ability to remove background noise, making it ideal for refining and polishing audio content.
- **Metadata editor**: This allows you to enter the texts for ID3 tags and add cover art directly to your output MP3 file from within the program.

• **Integration with Adobe Creative Cloud**: Audition integrates efficiently with Adobe Creative Cloud, allowing for seamless collaboration and the easy addition of mastered audio to other projects in, for example, Adobe Premiere Pro.

• **Free Trial**: Adobe offers a free trial of Audition, allowing users to explore its features and capabilities before making a purchase.

———

Neville Hobson

Social Strategist, Communicator, Writer, and Podcaster with a curiosity for tech and how people use it. Believer in an Internet for everyone. Early adopter (and leaver) and experimenter with social media. Occasional test pilot of shiny new objects. Avid tea drinker.

nevillehobson.com

CHAPTER 9
WEBSITES IN 2024

IN THE DYNAMICALLY CHANGING DIGITAL landscape of 2024, businesses must innovate to remain relevant. With the swift evolution of technology and shifting consumer behaviour, companies must adapt their website designs and functions to meet the new standards of user interaction and engagement. In this piece, we will dissect the pivotal trends defining web strategy and provide actionable insights to enable your business to leverage these changes proficiently.

Enhanced User Experience (UX)

Today's users demand frictionless digital interactions. Websites must be meticulously designed with intuitive navigation, swift load times, and personalised content at their core. The aim is not merely aesthetic appeal but a seamless user experience. Failure to prioritize UX can lead to diminished user engagement and lost conversion opportunities.

Mobile First Design

The ascendancy of mobile internet usage necessitates a mobile-first approach. A responsive design that renders effectively across devices, prioritising speed and accessibility, is mandatory. This approach is not just about scaling down a website, but rather optimising the entire experience to cater to the mobile user's needs.

Voice Search Optimization

Voice search optimization is now imperative. Accommodating voice queries and structuring content to be easily digestible by voice assistants is necessary to reach the widening user base that prefers speaking over typing.

Sustainability in Web Design

Sustainability is no longer a fringe consideration. Optimising your website to have a lower carbon footprint speaks volumes about your brand's values and consciousness. Employing energy-efficient coding practices, selecting green hosting providers, and minimising data transfer are essential steps towards a sustainable online presence.

AI and Chatbots

Entities are employing artificial intelligence to distil intelligence from data and chatbots to interact efficiently with users. Deploy these tools to provide instant customer service, tailored recommendations, and automate routine tasks, increasing efficiency and streamlining the customer's journey.

Interactive Content

Interactive elements such as quizzes, calculators, and 360-degree product views are becoming more prevalent. These tools engage users, provide them with valuable information, and can significantly boost time spent on site—beneficial for both SEO and user satisfaction.

Privacy-Centric Design

With the surge in data privacy concerns, transparency is paramount. Businesses must ensure their websites clearly communicate data practices and seek unequivocal consent where required. Privacy cannot be an afterthought; it must be embedded into the design process.

AR and VR Integration

Augmented Reality (AR) and Virtual Reality (VR) are transforming specific sectors like e-commerce and education. Integrate these technologies to offer immersive experiences that enhance user engagement and provide novel ways for users to interact with your products and services.

ACTION STEPS

1 Audit your website for UX optimization—streamline navigation, improve loading times, and personalise user content.

2 Ensure mobile-first design, focusing on the mobile user's speed and ease of use.

3 Adapt your content strategy for voice search—think conversational queries and succinct responses.

4 Review your code and hosting solutions for environmental impact and take corrective measures.

5 Implement AI and chatbots to enhance customer service and user engagement.

6 Create interactive content elements to enrich the user experience and entice prolonged engagement.

7 Strengthen your privacy policies and design for transparency and user consent.

8 Explore AR and VR potentials in your sector to offer breakthrough experiences on your site.

Leveraging these tactics, businesses can harness the full potential of their online presence, cater to modern users' expectations, and advance confidently in the competitive digital arena. The website of 2024 is not just a platform but a pivotal tool for business success, brand building, and customer connection.

The rapidity with which digital trends evolve necessitates vigilance and adaptability. Equip your website with the vital features and functionality that answer the needs of today and anticipate those of tomorrow. The companies that do so will carve a name for themselves in the annals of the digital age.

CHAPTER 10
YOUTUBE IN 2024

CHAPTER: **Social Media Strategies - Mastering YouTube for Business**

Introduction

In today's digital ecosystem, YouTube stands tall as a colossus of social media, offering vast potential for businesses looking to engage directly with their target audience. With over two billion users, YouTube's power extends beyond mere entertainment, establishing itself as an indispensable tool for compelling content marketing and audience engagement.

Red Bull's High-Flying Engagement

Red Bull's audacious "Red Bull Stratos" campaign encapsulates the profound impact YouTube can offer. An innovative combination of live streaming and thrilling content resulted in unprecedented visibility, illustrating the magnitude of creating real-time, captivating events that grip an audience's attention and foster a viral buzz.

Dove's Authenticity Resonance

Dove's "Real Beauty Sketches" serve as a powerful testament to content that strikes a chord emotionally. This campaign leveraged YouTube to not only disseminate a potent message but also to provoke discourse and deepen brand relationships on a scale traditional mediums seldom achieve.

Blendtec's Blended Virality

Blendtec, albeit with a more whimsical approach, demonstrated the power of product-centric content punctuated with a splash of humor. "Will It Blend?" transformed conventional product demonstration into a YouTube sensation, showcasing the platform's capacity to elevate brand profiles and convert viewership into quantifiable sales.

Nike's Cultural Dialogue

Nike's "Dream Crazy" campaign encapsulated advocacy and brand ethos simultaneously. Leveraging YouTube as a conduit for meaningful conversation, Nike validated YouTube's prowess in extending social commentary while bolstering brand standing, affirming it as an arena for more than product promotion – a space for principles.

Engaging the YouTube Audience

Effectively engaging a YouTube audience demands strategic planning and incisive execution. Here are several critical practices to consider:

1 Content Calendar Consistency: Regular, high-quality video postings are essential for nurturing a dedicated YouTube following. Construct a content calendar grounded in your brand narrative and strategic objectives.

2 Interactive Encouragements: Take advantage of YouTube's features to galvanize user interaction, deploying likes, comments, shares, polls, and community posts as standard mechanisms for dialogue.

3 Real-Time Connections: Live streams and premieres add a layer of immediacy, crafting moments for meaningful engagement and fostering a sense of community.

4 Direct Responses: Display appreciation for engagement by responding to user comments, further personalizing the user experience.

5 Strategic Collaborations: Partnering with other YouTubers can introduce your brand to diverse demographics, imbuing your content strategy with fresh perspectives.

6 Analytics-Driven Content: Utilize YouTube analytics to evaluate content efficacy, allowing you to refine your approach with precision.

CONCLUSION

YouTube offers an unparalleled platform for businesses wishing to connect with their audience in a direct and dynamic manner.

Through thoughtful strategy and consistent community engagement, businesses can expand their reach and foster enduring relationships with their target demographic. Inculcating these best practices into your YouTube strategy will not only enhance brand presence but also cement your status as a robust, responsive entity in the social media domain.

CHAPTER 11
TWITTER/X IN 2024

THE DIGITAL BUSINESS landscape is evolving at remarkable speeds, and at the forefront lies Twitter/X – a social media colossus that has undergone a momentous transformation. This chapter dissects the metamorphosis of Twitter/X into an instrumental platform for businesses and delineates strategies for harnessing its full potential as we venture into 2024.

THE EVOLUTION OF TWITTER INTO A POWERFUL INFORMATION AND ENGAGEMENT PLATFORM

Initially celebrated as a micro-blogging site, Twitter/X has burgeoned into a potent conduit for real-time information and user engagement. With a rich tapestry of features supporting multimedia content, live feedback, and hashtag-driven movements, Twitter/Xs utility transcends mere social networking.

Strategically leveraging Twitter/X starts with recognising its singular ability to propagate messages rapidly and its lever for influencing public discourse. Capitalising on this requires a deep understanding of the platform's mechanics and user behaviour—invaluable knowledge for businesses aiming to scale their influence and engagement in 2024.

LEVERAGING TWITTER/X FOR BUSINESS: UNDERSTANDING ITS UNIQUE VALUE PROPOSITION

To tap into Twitter/X's capabilities, businesses must first comprehend its unique value proposition. Unlike other platforms, Twitter/X excels in immediacy, brevity, and reach. In real-time, brands can resonate with global audiences, partake in trending conversations, and proliferate their messaging in concise, impactful bursts.

In the upcoming year, businesses should mould these attributes to their advantage by curating content that captivates and incites action, staying abreast of the zeitgeist, and becoming thought leaders within their domains.

NAVIGATING THE FUTURE OF TWITTER IN 2024: TRENDS AND PREDICTIONS

The future of Twitter/X is underscored by evolving trends and technological leaps. Anticipate greater integration of AI-driven analytics, augmented reality features augmenting user experience, and enhanced advertising options offering tailored outreach. Savvy businesses will remain vigilant to these developments, adopting innovations early to stay ahead of the curve.

Being proactive rather than reactive in incorporating new features and tools will be crucial in securing a competitive edge on Twitter by 2024.

STRATEGIES FOR MAXIMIZING BUSINESS VALUE ON TWITTER/X

The crux of maximising Twitter/X's business value lies in crafting coherent strategies aligned with overarching business goals. These strategies must be centred around increasing brand visibility, fostering robust customer relationships, and amplifying sales opportunities.

Deploy targeted content campaigns that speak to your demographic. Use advanced analytics to fine-tune your approach and make informed decisions about the type of content, the timing of posts, and the identification of influencer partnerships that can expand your reach.

THE ROLE OF DATA ANALYTICS AND PERSONALIZATION IN TWITTER ENGAGEMENT

Data analytics and personalization are pivotal to deepening user engagement on Twitter/X. By scrutinising data on user interactions, businesses can refine content strategies to resonate more profoundly with their audience, leading to a surge in brand loyalty and customer retention.

Invest in tools that provide insights into the performance of tweets and enable segmentation of your audience, thereby facilitating personalized communication that strikes a chord with different user groups.

BUILDING A STRONG BRAND PRESENCE AND COMMUNITY ENGAGEMENT ON TWITTER/X

Garnering a formidable brand presence on Twitter/X mandates consistency, relevance, and authenticity. Immerse your brand within the Twitter/X community by participating in relevant discussions, initiating campaigns that encourage user-generated content, and tapping into moments of cultural importance.

Community engagement springs from interaction. Accordingly, prioritise responsive dialogue, customer acknowledgments, and the fostering of a personable brand voice that aligns with your company values.

TWITTER/X AS A TOOL FOR CUSTOMER SERVICE AND FEEDBACK MANAGEMENT

Increasingly, consumers are turning to Twitter/X for real-time customer service and feedback. Businesses must equip themselves with robust support protocols, utilising Twitter/X to offer swift resolution to queries and grievances.

Offer clear and concise assistance, using automated responses for common queries and personalised attention for complex issues. Such attentiveness enhances customer experience and elevates your reputation as a client-focused brand.

COLLABORATION AND NETWORKING OPPORTUNITIES ON TWITTER/X FOR BUSINESSES

Twitter/X abounds with prospects for collaboration and networking. Identify synergistic businesses, thought leaders, and influencers with whom you can create joint ventures that propel shared interests and yield mutual benefits.

Such collaborations can take various forms, from co-hosted Twitter/X chats to cross-promotional posts, all serving to expand your audience and reinforce your business network.

THE IMPORTANCE OF AUTHENTICITY AND TIMELY ENGAGEMENT

Authenticity and timeliness are bedrocks of successful Twitter/X use. Consumers yearn for genuine interaction with brands and timely responses to global events and trends. Ensure your Twitter/X persona is reflective of your brand ethos, and demonstrate agility by adapting your content strategy in line with evolving conversations.

Prompt engagement not only amplifies your voice but also demonstrates your commitment to staying attuned to the pulse of society.

MEASURING SUCCESS: KPIS AND METRICS FOR TWITTER/X BUSINESS STRATEGIES

The application of measurable KPIs and metrics is non-negotiable when assessing the impact of your Twitter/X strategies. Track engagement rates, follower growth, click-through rates, conversion metrics, and brand sentiment to gauge your performance and orient future efforts.

Applying a meticulous framework for success measurement ensures accountability and enables continuous improvement in your Twitter/X undertakings.

FUTURE-PROOFING YOUR TWITTER BUSINESS STRATEGY: ADAPTING TO CHANGES AND INNOVATIONS

To future-proof your Twitter/X strategy, maintain a fluid and adaptable approach to accommodate the relentless pace of change. Stay educated on platform updates, regulatory shifts, and evolving user expectations, recalibrating your strategy as necessary to remain efficacious.

Consistently check the alignment of your Twitter/X initiatives with the broader vision of your brand. Through this lens, changes become opportunities rather than hurdles, as you sustain a dynamic, value-driven Twitter/X presence into 2024 and beyond.

CONCLUSION

In conclusion, Twitter/X's progression into an expansive platform of influential reach offers boundless potential for businesses. To capitalise on this, strategies must be insightful, nimble, and underpinned by a profound understanding of the platform's capabilities. Companies that can effectively harness the power of Twitter/X will differentiate themselves and achieve superior engagement, community-building, and business growth in the year ahead.

WHAT TO DO TO MAXIMISE YOUR ROI

• **Develop a Content Calendar:** Plan your tweets and content sharing to maintain a consistent presence, ensuring regular engagement with your audience.

• **Use Hashtags Wisely:** Implement relevant hashtags to increase the visibility of your tweets to a wider audience, but avoid overuse.

• **Engage in Trending Conversations:** Participate in relevant trending topics to showcase your brand's relevance and engage with a broader community.

• **Monitor Analytics Regularly:** Utilise Twitter's analytics tools to track your performance, understand your audience better, and refine strategies accordingly.

- **Leverage Twitter Ads:** Explore Twitter advertising to target specific demographics, increasing reach and engagement for important campaigns.
- **Focus on Customer Service:** Use Twitter as a platform for swift customer service response, enhancing customer satisfaction and brand loyalty.
- **Collaborate with Influencers:** Partner with influencers that align with your brand values to extend your reach within specific target audiences.
- **Showcase Brand Personality:** Ensure your brand's tone and personality are evident in every tweet, helping to build a unique brand identity.
- **Respond Promptly to Interactions:** Make it a priority to reply to comments, mentions, and direct messages to foster a sense of community and attentiveness.
- **Share Diverse Content Types:** Utilise various content formats like images, videos, polls, and GIFs to keep your feed vibrant and engaging.
- **Conduct Twitter Chats:** Host or participate in Twitter chats related to your industry to demonstrate thought leadership and engage in meaningful discussions.

CHAPTER 12
FACEBOOK PAGES
IN 2024

IN THE DIGITAL ERA, Facebook Pages have become a pivotal platform for businesses to build their online presence, engage with their audience, and drive commercial growth.

However, just being present on Facebook is not enough; companies must deploy strategic, well-planned practices to ensure they harvest the fullest benefits of this powerful medium.

This guide outlines a comprehensive action plan for businesses aiming to optimise their use of Facebook Pages alongside illustrative case studies that demonstrate both success and lessons from less effective approaches.

A ROBUST ACTION PLAN FOR FACEBOOK PAGE SUCCESS

Optimise for Engagement and Visibility

A Facebook Page must be well-crafted to capture attention. Ensure that all business information is accurate, incorporate a visually appealing cover photo, and use a clear, recognisable profile picture. The 'About' section should succinctly convey what the business offers, its history, and its mission. A call-to-action (CTA) button must be evident to facilitate customer action, such as 'Shop Now' or 'Contact Us'.

Commit to Quality Content

Quality content resonates with audiences and drives engagement. Diversify content types using images, videos, live streams, and polls to provide value and incite interaction. Post regularly and time releases for when the target audience is most active.

Harness Facebook Insights

Leverage Facebook's analytic tools to understand audience demographics and behaviour. Insights guide content strategies by showing which posts are garnering the most engagement and at what times the audience is most active.

Utilise Paid Promotions

Invest in Facebook ads to expand reach. Use the platform's robust targeting tools to specify audience characteristics such as age, interests, and location. Paid promotions can raise awareness, increase page likes, and drive specific actions.

Foster Community

Build a community by encouraging dialogue and engagement. Acknowledge comments, participate in conversations, and demonstrate responsiveness. Regular interaction builds trust and loyalty, driving word-of-mouth promotion.

Integrate with Other Channels

Facebook should be one facet of a broader marketing strategy. Share content across other social media platforms and link back to the Facebook Page. Use email marketing to alert subscribers to Facebook updates or promotions.

Monitor Feedback and Reviews

Engage with feedback publicly on the Facebook Page to show that the business values customer input and is committed to improvement. Take negative feedback offline when necessary to resolve issues.

Continuously Refine Strategy

What works today may not work tomorrow. Regularly review the performance of Facebook strategies and adjust based on data insights, feedback, and evolving digital trends.

CONCLUSION

Through a disciplined, well-informed, and responsive approach to managing their Facebook Pages, businesses can significantly amplify their online presence and build stronger relationships with their audiences.

By employing the strategies outlined—geared towards engagement, quality content, insightful analytics, targeted promotions, community building, integration, and adaptive learning—businesses can set themselves apart in the competitive digital marketplace, fostering sustainable, long-term growth.

Remember, success on Facebook doesn't come from happenstance; it's the product of strategic planning, execution, and ongoing refinement. Those who heed these principles stand to reap the most rewards from their Facebook Pages.

CHAPTER 13
INSTAGRAM IN 2024

INSTAGRAM HAS EMERGED as an essential tool for businesses to amplify their brand, connect with customers, and drive commercial success. Managing your Instagram account effectively is imperative to harness the full potential of this social platform. Here's a strategic action plan for small business owners, social media managers, and entrepreneurs to succeed with Instagram.

Introduction

Instagram is not just a visual platform but a dynamic marketplace and a community builder for businesses. With over a billion monthly active users, it offers commercial entities the opportunity to create impactful visual stories that resonate with their audience. Nonetheless, the efficacy of Instagram as a commercial tool hinges on strategic planning and adept management. This guide provides actionable steps to successfully manage your business's Instagram account.

STRATEGIC ACTION PLAN

1. Setting clear objectives

For Instagram management to bear fruit, delineating your key objectives is a prerequisite. Your objectives may encompass:
- Enhancing brand recognition within the target audience segment
- Increasing traffic to the business's online presence for heightened sales
- Building a community of active followers
- Bolstering the brand's industry authority
- Engaging with customers for support and brand loyalty
- Continually refining Instagram strategies based on data analytics

2. Choosing the optimal number of accounts

Though Instagram allows for the management of up to five accounts, for most small businesses, one account suffices to maintain focus and prevent audience dilution. However, if a business operates in distinctly different markets or offers contrasting product lines, additional accounts may be warranted. The decision should rest upon clear strategic needs and resources available for cohesive management.

3. Creating a consistent brand aesthetic

Your Instagram feed is your brand's storyboard. A consistent aesthetic is fundamental to establish brand identity. Choose a colour scheme, filter, and layout that aligns with your branding guidelines. Regularly using templates branded with your logo, font, and colour palette fortifies brand recall among followers.

4. Crafting high-quality content

Content is king on Instagram. Crafting high-quality images and videos is non-negotiable for engaging audiences. Regardless of the product or service, visual storytelling can captivate attention, evoke emotions, and drive user action. Invest in good photography equipment or professional services to ensure your content stands out.

· · ·

5. Scheduling content strategically

Leverage tools like Hootsuite or Later to schedule your posts for optimal times when your audience is most active. A consistent posting schedule keeps your brand at the forefront of your followers' minds without overwhelming them with too many posts.

6. Utilising Instagram's range of features

Instagram offers a plethora of features such as Stories, IGTV, Reels, and Shopping. Each serves unique purposes:

• **Stories**: Ideal for sharing behind-the-scenes content, limited-time offers, or direct follower engagement through polls and Q&A sessions.

• **IGTV and Reels**: Offer longer-form content opportunities and bite-sized videos, respectively. They are perfect for tutorials, product demos, and sharing brand narratives in an engaging format.

• **Shopping**: Facilitates in-app sales enabling users to purchase without leaving the platform. It is particularly lucrative for product-based businesses to streamline the shopping experience.

7. Engaging with the community

Engagement is a two-way street on Instagram. Responding to comments, engaging with followers' content, hosting Q&A sessions, and replying to direct messages fosters a sense of community. Utilise engagement metrics to understand what content resonates best with your audience, which can inform future content creation.

8. Collaborating with influencers and brands

Networking with influencers and related brands can amplify your reach. Collaborations and sponsorships should align with your brand values and appeal to your target demographic for brand authenticity and credibility.

9. Using hashtags wisely

Strategically using hashtags boosts your content's discoverability. Research trending and industry-specific hashtags to reach a broader audience. Custom branded hashtags can also encourage user-generated content and increase brand engagement.

10. Analysing and optimising strategies

Use Instagram analytics to measure the performance of your content and strategies. This could inform adjustments in your content strategy, posting schedule, and engagement practices.

CONCLUSION

Masterful management of your Instagram account requires a well-crafted blend of strategy, creativity, and community engagement. Focusing on clear objectives, cultivating a consistent brand aesthetic, publishing exceptional content, engaging with followers sincerely, and perpetually optimising your strategies are the cornerstones of success on this platform.

By implementing the outlined actions diligently, with a magnetic brand narrative and compelling visual content, your Instagram account will not only flourish in follower numbers but in fostering genuine brand-consumer relationships and achieving tangible business outcomes.

CHAPTER 14
TIKTOK IN 2024

TIKTOK HAS RAPIDLY BECOME a social media juggernaut, capturing the imagination and attention of users across the globe. With its unique algorithm and engaging content format, TikTok presents an unparalleled opportunity for businesses to increase their visibility and connect with their audience on a more personal level. Yet, with great potential comes the need for a clear and robust strategy.

In this chapter, we will explore how businesses can leverage TikTok effectively and lay out a definitive action plan for success.

CRAFTING YOUR TIKTOK STRATEGY

Understanding TikTok's Ecosystem

Before a business can truly harness TikTok's potential, it's crucial to understand the platform's nature. TikTok thrives on creativity, spontaneity, and trends. It's a space where the audience yearns for authenticity and engagement through short, entertaining videos. The platform continually evolves, so staying updated with its trends is essential.

Positioning Your Brand

Businesses must position themselves in a manner that aligns with the TikTok community. This involves crafting a brand persona that resonates with TikTok users while remaining true to the company's core values and messaging. Through a mix of market analysis and creative brainstorming, businesses can find their unique voice on the platform.

Content Creation and Curation

The heart of TikTok is its content. Businesses must focus on creating and curating content that not only represents their brand but also appeals to the TikTok audience. This content needs to be consistent, of high quality, and most importantly, engaging. The integration of popular trends, effects, and music can amplify your content's reach.

Measuring Success

Any strategy is incomplete without a framework for measurement. Key performance indicators (KPIs) on TikTok include engagement rates, follower count, video views, and user-generated content.

Analytics tools provided by TikTok can help track these metrics and inform the refinement of your strategy over time.

TOP 10 THINGS TO DO TO WIN WITH TIKTOK

1 Align Content with Objectives: Tailor your content to embody your brand and meet your set objectives - whether it's driving traffic to your website or increasing brand recognition.

2 Leverage Hashtags: Utilize trending and relevant hashtags to increase the visibility of your content in TikTok's search function.

3 Engage with Trends: Participate in challenges and trends while putting your unique spin on them to maintain brand relevance.

4 Stay Consistent: Post regularly to maintain visibility and keep your audience engaged with fresh content.

5 Create High-Quality Videos: Invest in good production quality to make your content stand out.

6 Utilize TikTok Ads: Take advantage of TikTok's advertising platform to target specific audience segments and drive tangible results.

7 Engage with Your Community: Respond to comments, like videos, and follow back to build a dedicated community around your brand.

8 Collaborate with Influencers: Partner with TikTok influencers who resonate with your brand to tap into their follower base.

9 Track and Analyze Performance: Regularly review analytics to understand what works and refine your strategy accordingly.

10 Offer Value: Whether it's entertainment, information, or discounts, ensure your content provides value to your viewers, encouraging them to return.

AN ACTION PLAN FOR SUCCEEDING ON TIKTOK

To manage and succeed with TikTok as a business, here is a step-by-step action plan:

1 Define Clear Goals: Establish what you want to achieve with your TikTok presence. Goals should be specific, measurable, attainable, relevant, and time-bound (SMART).

2 Know Your Audience: Conduct research to grasp the demographics, preferences, and habits of your target audience on TikTok.

3 Develop a Content Calendar: Plan your content in advance, including campaigns for product launches or special events.

4 Optimize Posting Times: Determine the best times to post on TikTok when your audience is most active for maximum engagement.

5 Monitor Trends: Keep an eye on TikTok trends and think creatively about how your business can participate or create similar viral content.

6 Evaluate and Iterate: Regularly review your TikTok analytics to understand which types of content perform best and adjust your strategy accordingly.

7 Invest in Your Community: Foster relationships by engaging with users and creating a sense of belonging around your brand.

8 Leverage User-Generated Content: Encourage and share content created by your followers that feature your brand, as it can boost authenticity and reach.

9 Stay Compliant: Ensure all content adheres to TikTok's community guidelines and copyright laws.

10 Educate Your Team: Keep your team informed about best practices, platform updates, and content guidelines to maintain a cohesive and effective TikTok presence.

CONCLUSION

In conclusion, the future of TikTok lies in the hands of brands who are willing to adapt, listen, and engage with their audience on a personal level. By following a methodical, yet flexible approach to content creation, community building, and trend adaptation, businesses can achieve maximum results from their TikTok accounts. Employing the top 10 winning strategies and following the action plan will provide a strong footing for businesses venturing into TikTok's dynamic terrain.

CHAPTER 15
NEWSLETTERS IN 2024

THE EVOLUTION OF NEWSLETTERS: A BUSINESS GUIDE TO THE NEXT THREE YEARS

IN THE TUSSLE for customer attention within the digital sphere, newsletters have secured their place as an invaluable tool in a business's marketing arsenal. With the landscape constantly evolving through technology and user behaviour, businesses must anticipate and adapt to stay apace.

Over the next three years, I anticipate revolutionary shifts brought on by technological advancements and ever-changing customer expectations. Navigating these changes will require a blend of technical know-how and astutely crafted strategies. Below, I present a comprehensive guide on the future of newsletters and methods to ensure your business maintains, if not enhances, its engagement with customers.

Personalisation and AI

The era of generic newsletters is waning. Precision targeting through Artificial Intelligence (AI) will render newsletters more significant to individual needs and interests, fostering a more intimate connection between brand and consumer.

. . .

Mobile-First Tactics

With mobile devices now the primary interface for email engagement, newsletter design will see a pivotal shift towards mobile optimisation, ensuring visual allure and content relevance on smaller screens.

The Rise of Interactive Content

The passive consumption of content is on the decline; the next wave will favour newsletters that invite readers to interact through polls, videos, and quizzes, metamorphosing static content into a dynamic exchange.

Automated Efficiency

The future holds a more strategic yet effortless approach to newsletter campaigns, with automation enabling real-time, behaviour-triggered communication that aligns with users' immediate context and interests.

Privacy as Priority

In an era marked by heightened sensitivities around privacy, newsletters that secure consent and uphold data integrity will foster trust and sustained readership.

Sophisticated Segmentation

Expect granular analytics and segmentation strategies to burgeon, giving newsletters a more nimble form, adept at showcasing tailored content to specific user groups.

New Frontiers: Voice and Wearables

With voice recognition technology and wearable devices carving new territories in user engagement, newsletters must adapt, venturing beyond the written word into more accessible and instant notification systems.

Eco-conscious Messaging

Sustainability will inform not only the content but also the operational ethos behind newsletters. Green hosting and energy-efficient design will resonate with the ecologically-conscious subscriber.

Community Cultivation

Newsletters will transcend their traditional function, knitting communities by enabling conversations and partnerships amongst readers, and thus, deepening brand loyalty.

Data-Driven Decisions

Analytics will underpin newsletter strategy, with performance metrics acting as compasses guiding content creation, frequency, and user experience enhancements.

This trajectory predicates a wealth of opportunities but requires tenacity to harness these burgeoning capabilities effectively. The subsequent section encapsulates a distilled set of principles to aid in the composition of successful newsletters.

TOP TEN PRINCIPLES FOR NEWSLETTER SUCCESS

1 Personalize Profoundly: Ensure content resonates on a personal level using AI-derived insights.

2 Engage Creatively: Incorporate captivating subject lines and a diversity of content formats. Hint: there's a few AI engines that are both creative with headlines and can take an existing headline and reply with ten headlines. Some of the headlines might be less that desirable, even clickbait, but some will be better than you can write—and the AI engine takes about two seconds to write all ten for you to choose from. I use Jasper.ai and I'm very happy with it for all business writing.

3 Optimise for Mobile: Design with a mobile-first philosophy to capture the growing on-the-move audience.

4 Automate Intelligently: Use behaviour-triggered emails to communicate with relevance and timeliness.

5 Guard Privacy Fervently: Treat data with integrity, adhering to stringent security and privacy standards.

6 Segment with Finesse: Craft content that speaks to segmented audiences with distinct interests and preferences.

7 Innovate Continuously: Experiment with new technologies and platforms to enrich the reader's experience.

8 Conduct with Sustainability: Apply sustainable practices to newsletter campaigns showing responsibility towards the environment.

9 Forge a Community: Employ newsletters to build and nurture a community around shared values and interests.

10 Analyse and Adapt: Keep a vigilant eye on metrics and be prepared to revise strategies responsively.

I now turn to a practical roadmap that charts the process of launching and developing a thriving newsletter service.

TEN-STEP GUIDE TO LAUNCHING AND GROWING A NEWSLETTER

1 Define Your Value Proposition: Clarify the unique content or service your newsletter offers.

2 Build a Robust Platform: Select a reliable email marketing tool that offers analytic capabilities, such as Mailchimp or Sendinblue. Avoid any small-time tools, perhaps that come with your website hosting package, because they will likely be marked as spammers by Google and AI email engines. For your readers and your business's sake, stick to the major players in this part of the business use of social media; your stakeholders will thank you for it.

'Fast, cheap, and good... pick two. If it's fast and cheap, it won't be good. If it's cheap and good, it won't be fast. If it's fast and good, it won't be cheap.'

—Jim Jarmusch

3 Craft Your Subscription Process: Develop a straightforward and respectful sign-up mechanism.

4 Design with Clarity and Impact: Prioritise readability and engaging layouts that work across devices.

5 Create Engaging and Relevant Content: Focus on quality content that aligns with your brand's expertise.

6 Strategise Your Sending Schedule: Determine optimal timings and frequency for your emails.

7 Execute and Automate Campaigns: Implement triggered emails for contextual relevance.

8 Grow Your Subscriber List Ethically: Promote your newsletter across platforms and encourage sign-ups without resorting to coercive tactics.

9 Analyse Feedback and Metrics: Use tools like Google Analytics to understand subscriber behaviour and preferences.

10 Refine and Expand: Continually refine content based on feedback and explore new interactive elements for expansion.

The coming years promise a kaleidoscopic shift in the newsletter landscape—one which businesses must approach with both fervency and finesse. It is incumbent upon management to digest these insights and inculcate them into actionable and strategic planning, fostering not just adaptation, but the very evolution of newsletters at their charge.

CHAPTER 16
CRISIS MANAGEMENT

THE RELIANCE ON SOCIAL MEDIA FOR CRISIS MANAGEMENT

CRISES CAN STRIKE a business at any time, and when they do, the difference between a hiccup and a full-blown catastrophe often hinges on an organization's ability to swiftly communicate, respond, and manage the situation effectively. Social media has become an indispensable tool in the crisis management arsenal, offering unparalleled speed, reach, and direct engagement with stakeholders during times of turmoil.

CRISIS MANAGEMENT: A DEFINITIVE NEED IN BUSINESS

Crisis and Consequence

Crisis management, fundamentally, refers to the processes and protocols an organization implements to handle unforeseen, adverse events that can jeopardize its integrity, reputation, or existence. The outcomes of these incidents can range from minor operational setbacks to existential threats with severe economic and societal impacts.

The Imperative for Preparedness

Expert Melissa Agnes emphasizes that, "Crisis readiness is not about the 'if,' it's about the 'when.'" This perspective underlines the significance of maintaining a comprehensive crisis management strategy, anticipating various forms of jeopardy such as reputational, operational, or managerial crises.

SOCIAL MEDIA AS A CRITICAL CRISIS MANAGEMENT TOOL

Digital Nerve Centre

Social media platforms function as a digital nerve center for public sentiment, enabling businesses to monitor, analyze, and respond to stakeholder concerns with immediacy. Their widespread adoption confers upon enterprises the ability to disseminate information rapidly across a range of demographics and geographies.

Swift and Scalable Communication

The vitality of social media in crisis communication is tied to its capacity for swift message propagation, thereby escalating the velocity at which information spreads and opinions are formed. Professor Scott Galloway expounds on this, stating that "In a crisis, social media channels are the bloodstream of information distribution."

THE DOUBLE-EDGED SWORD OF SOCIAL MEDIA

Viral Velocity

While speed is of the essence in crisis management, it can also be a double-edged sword. Social media facilitates the rapid dissemination of information, which includes misinformation or speculation that can exacerbate a crisis. Sailing these treacherous waters demands that businesses engage promptly with their stakeholders to correct inaccuracies and provide a voice of reason and clarity.

Reputational Risk and Public Backlash

Understanding the potential for reputational risk and public backlash on social media is crucial. A misstep or misguided post can provoke immediate and often impassioned public feedback. Therefore, the tone and content of communications must be meticulously managed to align with public expectations and sentiments.

I and many of the early adopters of social media will not have forgotten the fate of Kryptonite, a small business that made a bike lock aimed at cyclist messengers. These cyclists quite often entered into buildings to deliver parcels, only to exit the building to find their bike had been stolen.

Kryptonite boasted that their expensive premium lock was unbreakable, and that they would give a handsome reward to whoever proved it otherwise. Many took up the challenge and failed. Kryptonite believed themselves and their product invincible.

Then one day in 2004 bike enthusiast and network security consultant Chris Brennan posted in a forum that in no uncertain terms the bike lock **could be broken**. And not with some fancy hi-tech equipment, but with a **Bic biro**. [**https://youtu.be/LahDQ2ZQ3e0?si=mt-BXZ1V7tfyV44fL** is one of many videos made around that time that showed the world how to break an unbreakable bike lock]

The blogosphere, as it was known, went into meltdown. Kryptonite the business was flooded with enquiries, sarcasm, vitriol, and more. They were taken by surprise, understandably, and the small company of just 25 worked around the clock to fix the fault and once more be able to boast about the unbreakability of their lock.

But because of 'all hands on deck', their communications star Donna Tocci went 'radio silent'. No communications went out during this period.

This infuriated the blogosphere and suddenly the event became of mythical proportions. It seemed like the whole social media world had an opinion about both the incident and the company.

> When a company's not communicating with its market and existing customers, imaginations run away with themselves and only the worst becomes the default position of the market and audience.

BEST PRACTICES FOR CRISIS MANAGEMENT ON SOCIAL MEDIA

Transparency and Timeliness

Transparency and timeliness are keystones in utilizing social media effectively during crises. To echo Simon Sinek's principle, "Transparency doesn't mean sharing every detail—it means honesty and openness." Establishing a reputation for transparent communications before a crisis occurs can significantly enhance trust during high-pressure situations.

Strategies for Social Media Engagement

Drafting a crisis communication plan with predefined protocols, potential scenarios, and designated spokespeople streamlines the decision-making process and ensures a coherent message. The plan should stipulate how to adapt messaging to the evolving dynamics of a crisis, always adhering to a sense of urgency and accuracy.

THE PITFALLS OF CRISIS MANAGEMENT VIA SOCIAL MEDIA

The Misinformation Quagmire

Misinformation is an omnipresent risk when managing crises on social media. A well-prepared organization employs advanced social listening tools to detect and counter false narratives, maintaining control over the storyline as it unfolds.

Managing Emotional Responses

Social media, inherently personal and emotionally charged, can amplify the emotional intensity of a crisis. Preparation, professionalism, and emotional intelligence are essential to navigating these challenging dynamics with sensitivity and composure.

LEARNING FROM EXPERIENCE: CASE STUDIES IN SOCIAL MEDIA CRISIS MANAGEMENT

Success Stories

There are innumerable instances of businesses leveraging social media to transform potential disaster into a demonstration of resilience and reliability. These successes often share common traits, such as rapid response, empathic communication, and unwavering commitment to resolving stakeholders' concerns.

Cautionary Tales

Similarly, there are cautionary tales of inadequate social media crisis management. Instances where a lack of prompt acknowledgment, defensiveness, or insincerity has resulted in long-term reputational damage illustrate the importance of a well-considered strategic approach.

Preparing for the Inevitable: Crisis Communication Plans

Every business susceptible to a crisis must have a meticulous crisis communication plan, which includes extensive training on social media use during emergencies. Regular simulations and updates ensure that the organization remains agile and prepared for any crisis that may arise.

ANTICIPATING THE FUTURE OF SOCIAL MEDIA IN CRISIS MANAGEMENT

The Advent of AI and Mobile-First Strategies

Looking ahead, the utilization of AI and machine learning for early crisis detection will become increasingly prevalent. A mobile-first communication approach reflects the growing trend where the public turns to smartphones as their primary source of news and information.

The Emergence of Video and Influencer Engagement

The ascendancy of video as a medium for crisis communication underpins the demand for authenticity, while collaboration with influencers can effectively reach niche audiences. Profound engagement both prepares and protects a brand's reputation.

Navigating Evolving Platforms and User Behaviour

Businesses must stay dynamic, evolving their crisis response tactics as social media platforms and user behaviours change. Cultivating credibility through consistent, genuine interaction establishes a bulwark against potential crises.

Privacy and Data Security

Finally, privacy and data security are becoming cardinal concerns for social media communications during a crisis. It is imperative for businesses to handle customers' personal information with respect for the intricate tapestry of data protection regulations.

CONCLUSION

Harnessing social media's power for crisis management demands a nuanced understanding of the platform's inherent advantages and pitfalls. By crafting tailored strategies and maintaining a commitment to transparency, integrity, and engagement, businesses can transform potential catastrophes into opportunities to fortify trust and loyalty with their audience.

Thus, in an era where digital readiness can determine an organization's survival, a robust strategy for managing crises on social media is not merely prudent—it's paramount.

CHAPTER 17

PODCASTING TRENDS AND THE FUTURE OF AUDIO ENGAGEMENT

THE PODCASTING medium has rapidly evolved from a niche form of media consumed by a dedicated few to a sprawling industry with mass appeal. Businesses, content creators, and advertisers have taken notice, capitalising on the personal and direct relationship that podcasting facilitates between them and their audience. This chapter will succinctly dissect the future trajectory of podcasting, assess which businesses are gaining traction through audio content, and provide practical insights into why companies should—or shouldn't—add podcasting to their communication strategies.

Interactive Podcasting and the Rise of Listener Engagement

The traditional podcast format is being revolutionised by interactive elements that turn passive listeners into active participants. Future predictions point towards burgeoning interactivity within podcasts, stretching from direct Q&A sessions to real-time polling. This immersion enhances the listener's agency, making the content more relevant and engaging while providing valuable feedback to the creator.

Niche Podcasting: Specialised Content for Targeted Audiences

The specificity that niche podcasting affords cannot be overstated. Creators are continually honing their subject matter, addressing the minutiae of interests preferred by their listeners. This hyper-targeting allows businesses to become authoritative voices within their domains, demonstrating knowledge and adding tangible value to the listener's experience.

Short-form Content: Brevity Meets Demand

In the flux of content consumption preferences, brevity is king. Short-form content aligns with the demanding schedules and fragmented attention spans of modern listeners. Businesses and creators are expected to distil valuable content into shorter episodes, delivering impactful messages that resonate quickly and memorably.

Case Studies of Successful Business Podcasts

Several prominent companies exemplify the power and potential of podcasting as a marketing and engagement tool:

Starbucks' "Upstanders" Series

This series complements Starbucks' ethos of social responsibility and community support, strengthening brand identity while engaging customers at a deeper level.

Mailchimp's "Mailchimp Presents" Platform

Mailchimp extends its brand reach beyond email campaigns by delivering entrepreneurship-focused podcasts, showcasing storytelling excellence in content marketing.

HubSpot's "The Growth Show"

By offering insight into business growth, HubSpot emerges as a thought leader in the realms of marketing and sales, reinforcing its market standing.

Shopify's "Start Yours" Podcast

This resourceful podcast for entrepreneurs mirrors the support and guidance Shopify provides, fostering a symbiotic relationship with its e-commerce platform.

Mastercard's "Fortune Favors the Bold"

The series affirms Mastercard's position as an innovator within the financial landscape, through candid discussions about the future of business and technology.

INTEGRATING PODCASTING INTO A COMMUNICATION STRATEGY

Podcasting serves as a complement to existing marketing content. Integrating this auditory form into a broader communication mix adds depth and variety. It is a versatile medium that supports content marketing, enhances brand recognition, and helps in building a cohesive and comprehensive narrative. Podcasts can introduce brand values, share success stories, and disseminate thought leadership, potentially fostering a community around shared themes.

The Business Benefits of Podcasting

The merits of adopting podcasting as part of business strategies are manifold:

• Podcasts augment visibility and brand awareness.

• They offer a platform to establish thought leadership in a specialised niche.

• This format encourages community building around a brand.

• Diverse content marketing strategies benefit from the addition of podcasts.

• They have the potential to generate new leads.

• They can also serve as a communication tool within a company.

• Analytics from podcast content provide actionable insights.

• They offer new revenue streams through sponsorship and premium content.

• Podcasting adapts to the vagaries of consumer behaviour and preference shifts.

However, before equipment purchasing and content development begins, assessing the alignment of podcasting with business objectives is critical.

Equipment Essentials for Professional-Sounding Podcasts

The baseline for professional podcast production is clear audio quality, which necessitates specific equipment:

• A high-quality microphone is non-negotiable for crisp, clear sound.

• Headphones are essential for monitoring audio levels and quality during recording.

• Recording software for capturing and editing content is a necessity.

• A pop filter can minimise plosives for a smoother sound profile.

• Acoustic treatment for the recording area is beneficial for reducing echo.

Striking a balance between quality and budget is crucial, as is ensuring that investment matches the strategic commitment to podcasting as a medium.

Reasons Against Podcasting for Some Businesses

Not every business is suited to podcasting. Resource constraints can curtail the production of high-quality content. The uncertainty of direct ROI might deter profit-driven organisations. Additionally, the saturated podcast market can make listener acquisition and retention an insurmountable hurdle for some. Lack of in-house expertise and potential legal risks are further deterrents. Hence, there must be a compelling rationale aligned with long-term business goals for podcasting to be pursued.

CONCLUSION

The podcasting sphere continues to blossom, with interactive podcasts, niche content, and short-form audio posited to dominate the domain. Businesses can leverage podcasts to enhance brand identity, cement thought leadership, and galvanise community engagement. The correct equipment arsenal ensures quality production, but ventures into podcasting must be cautiously weighed against the potential detriments and strategic fit.

Podcasting, executed with due consideration and skill, can add a resonating voice to the chorus of a brand's communication efforts, fostering deep connections in ways few other mediums can. However, the considerations of resource allocation, content value, and strategic alignment necessitate a grounded and systematic approach to integrating podcasts into business strategies, ensuring that each audio episode advances the overarching narrative a business seeks to cultivate.

CHAPTER 18
PODCASTING STRATEGY

IN THE EVOLVING landscape of digital marketing, podcasting stands out as a powerful medium for storytelling, sharing insights, and building communities. By 2024, businesses that harness this platform effectively can achieve significant engagement and brand loyalty. This guide equips enterprises with a robust strategy to master the medium and captivate their audience.

Crafting a Compelling Content Strategy

To utilise podcasting effectively, a meticulously planned content strategy is essential. It should align with your business's overarching aims while catering to your audience's preferences. Start with identifying the USPs (Unique Selling Points) you wish to communicate and the topics that will resonate most with your listeners.

Podcast Series and Episode Planning

Initiate with a thematic series that encapsulates your brand's expertise or unique perspective. Plan episodes that break down the overarching theme into digestible and engaging content segments tailored for easy listening.

Audience Research and Engagement

Conduct thorough research to decipher your audience's pain points, desires, and habits. Regularly invite feedback and engage in dialogue to finesse your content based on listener preferences and industry trends.

Consistency and Quality

Ensure regular episode releases to establish predictability and form a habit with your audience. Prioritise quality content that uplifts, educates, or entertains to provide enduring value.

SOCIAL MEDIA STRATEGY FOR PODCAST PROMOTION

Social media channels are potent tools for amplifying your podcast reach. Employ a multi-tiered social media approach to not only announce new episodes but also to entice and engage with your audience.

Teaser Content

Create buzz with teaser clips, quotes, or infographics that capture your upcoming episode's essence in seconds.

Interactive Elements

Foster community through polls, AMAs (Ask Me Anything), and other interactive content that links directly back to your podcast.

Incentives and Lead Magnets

Provide listeners with compelling reasons to tune in, such as exclusive content or downloadable resources tied to your episode's theme.

PRODUCING A PROFESSIONAL PODCAST

The production quality of your podcast can set you apart. Focus on clarity, seamlessness, and professionalism in every audio file released.

Invest in Quality Equipment

At minimum, invest in a high-quality microphone, headphones, and audio interface. Consider the acoustic environment, seeking a quiet, soft-furnished space to minimise echoes.

Editing and Post-Production

Use a reputable DAW (Digital Audio Workstation) to edit your recordings. Trim unnecessary segments, level out audio, and integrate transitions seamlessly. For those less technically inclined, outsourcing to a professional editor is a viable option.

I've used the full-featured and free editor 'Audacity', but I've also used Adobe's 'Audition', which while not free, is a top-tier editor with a beautiful interface.

Hosting Multiple Presenters

When hosting multiple presenters, utilise multi-track recording for control over each voice in post-production. Quality call-recording software or a cloud-based recording solution can be employed for remote interviews.

Selecting a Hosting Platform

Choosing the right hosting platform for your podcast extends your reach and simplifies distribution. Consider platforms that offer robust analytics, ease of distribution to multiple platforms, and reliable uptime. Anchor, Podbean, and Buzzsprout are notable options offering scalable solutions. I've used Libsyn since the mid-2000s and am very happy with their offering.

Sourcing Music and Sound Effects

A podcast's auditory experience can be greatly enhanced by music and sound effects. Use royalty-free libraries or purchase licenses from platforms such as Epidemic Sound, AudioJungle, or Soundstripe for access to a wide selection of high-quality tracks and sounds.

———

CONCLUSION

This comprehensive guide articulates a strategy with precision and authority, providing actionable steps for businesses to excel in the podcasting arena of 2024. By employing these recommendations, your business can produce podcasts that deliver compelling content, enhance brand presence, and foster community engagement.

CHAPTER 19
SCRIPTING A PODCAST

DONNA PAPACOSTA

One of the most popular questions I hear from novice and would-be podcasters is: "Should I script my podcast?"

Queries related to podcast scripting are also the most common keyword searches on my blog.

I've written this book to help answer these questions, so you can benefit from my many mistakes. (Lesson in humility: listen to your first five podcast episodes.)

Of course, the answer to the podcasting scripting question is not a straight "yes" or "no." It's more like: "It depends."

Let's start with the positives and negatives of scripting.

CONVERSATIONAL PODCASTS DON'T NEED A SCRIPT

If your podcast consists of interviews or conversations between you and one or more guests, I'd suggest not using a script at all. Instead, write out a list of topics you want to discuss with your guest, and share this list with him. However, do not send the precise questions you're planning to ask. Why not? Because some guests, believe it or not, will scribble out their exact answers and then expect to *read* them verbatim during the recording session. This is especially true when someone is nervous and has little experience with interviews.

In my early days of podcasting, I wanted to be nice, so I would send people questions in advance when they asked for them. As a result, I had to stop people dead cold in the middle of one of their answers to say: "Wait a minute. You sound a bit stilted. Are you reading? Please don't. Let's have a conversation instead. You'll thank me later." And they did.

Needless to say, I no longer email exact questions to guests in advance – just general question areas.

Why am I so against this type of reading? Unless you're a trained voice actor, you're probably not very convincing reading your answers. In fact, it's likely that you'll read in a monotone and bore people to death. Trust me: *They can tell you are reading*.

The structure of your language and the tone of your voice are giveaways.

As the podcast host, if you are afraid of forgetting something, by all means write your main points down as a checklist for yourself. This includes the questions for your podcast guest. Keep in mind that your podcast will most likely be edited. In the worst case, you can record and add forgotten content later. So, don't worry about leaving something out.

In my experience, the best interviews happen when I start off with questions that are merely guideposts, and let the conversation branch off naturally, with an easy flow. Don't worry if you ask only five of your seven planned questions, as long as the content is good, and your listeners will benefit from hearing the conversation.

INTERVIEW TIPS FOR PODCASTS

If you are recording your podcast interview in person, you have a bit of an advantage, I think. You and your guest can see each other and read each other's body language. Most podcast interviews, however, are recorded remotely. (I use Skype and Call Recorder on my MacBook Pro most of the time.)

Let's imagine you are in Chicago and your guest is in Paris. You have sent her question areas, not exact questions in advance, and she knows what the goals of the interview are. Before you start recording, try to set her at ease. Sometimes a podcast guest is quite jittery at the start, and you'll do everyone a favour if you can help her relax. Believe it or not, I've interviewed CEOs who were quaking in their boots.

To counter the nerves, chat about the weather in Paris. Ask about her kids or her dogs or her latest project. Don't talk too much about the content you're going to discuss in the podcast, however, because you want it to be fresh while you're recording.

Before you begin the formal interview, remind your guest that you won't be uttering agreements as she talks. In other words, you will avoid interjecting such phrases as "Oh, I see" or "Uh huh" while she speaks, as you normally would during an in- person conversation. (Well, *I* normally would. One of the drawbacks of growing up in New York is my tendency to want to jump in when someone speaks. I don't like to think of this New York trait as interrupting *per se*; I'm adding value!)

Also, do tell your guest you will leave a pause when you think she has finished saying something. By the same token, you want her to pause when *you* have finished speaking.

Without the visual cues, it's hard sometimes to know when someone is actually done making a point, or if they're just pausing momentarily for emphasis or to gather their thoughts. You want to avoid your voices jumping on top of each other. It's simple to edit out dead air. It's harder to edit two voices laid on top of each other.

Bonus: Before a first interview with someone, unless they're experienced with podcasts, I usually send them a PDF tip sheet called "How to be a terrific podcast guest." Here it is for you below. Feel free to customize it and share it with your own guests.

HOW TO BE A TERRIFIC PODCAST GUEST

Since 2005 I've conducted hundreds of podcast interviews, many in person but most remotely. Various factors affect the success of a podcast, both from a content and audio quality point of view. But you, as the interviewee, are the most important.

Here are a few tips to help you shine as a podcast guest. After all, you're probably doing the interview to share your knowledge as an expert, to promote your business or your book, or to generate speaking gigs. So, you want to be your best.

Preparation

Check out the podcast to ensure it fits your needs. Ask the host what the angle of the interview is, and about the intended audience. If your target market is mega- corporations, and the podcast is geared toward home business, you should probably decline.

Ask for discussion areas in advance, but don't expect to get every question in writing. As a podcaster, I don't share exact questions before the interview, because some guests get a little nervous and script their answers, and then expect to *read* them on the podcast. Trust me on this: Unless you're a trained voice actor, reading can be the fastest way to ruin an interview. Of course, you do want to keep relevant facts and figures at your fingertips so you sound like the expert you are.

Get the technology out of the way early. When you're booked as a guest, confirm the date and time (including time zone). Ask whether the interview will be conducted in person, by phone or via Zoom.

SOUNDING GREAT

These days, the majority of podcasters will use Zoom for remote interviews. Although the popularity of Zoom continues to grow, some guests are not familiar with it, and attempt to "try it out" at the time of the interview. This is not a good idea. If you're not accustomed to using Zoom, install it in advance and experiment with a friend or family member. Learn how to start up the program, plug in your headset and/or microphone, and turn off notifications so you don't hear annoying beeps during the interview.

The right gear will improve sound quality immensely. If you're going to do multiple remote interviews, invest in a headset and microphone combo. Even better, a USB microphone would be a step above a one-piece headset/mic combo. (The Blue Yeti gets high ratings.) If you don't have proper headphones, you can use the earbuds from your smartphone. Please avoid using the speakers on your computer for output. The worst audio quality will result from using the microphone built into your PC. If you must rely on it, learn to use it properly and position yourself for optimal sound.

Quiet on the set! If you have a thunderous fan or air conditioner in the room, turn it off for the interview. If you have a noisy child or pet, do not do the interview if they're within earshot, unless they're part of the story.

Share your stories. Using a story to make a point can be very effective. But do be sure your stories don't run too long. Check with the podcaster: how much time do you have? Remember not to talk too fast. Sometimes when we're excited about a topic, we speak too quickly. Slow down, and remember to breathe.

Pauses please. The interviewer might ask for pauses between questions and answers, which makes the editing process easier. So, after you say something, don't be afraid of the silence, which will be edited out. When I do interviews, I usually warn the guest in advance about these pregnant pauses.

Yes, sometimes you must script

Many times I've talked a client out of scripting a podcast for the reasons I've outlined, but sometimes you have to bite the bullet and let the scripting begin.

Fellow podcaster Steve Lubetkin, commenting on a blog post of mine about podcast scripting says: "We script intros and outros, and for some clients in heavily regulated industries, we prepare scripts in advance for legal review. It's a lot easier to get lawyers to approve a script in advance and stick to it than to have them picking apart the recording afterward and having to re-assemble the guests and re-record the interviews."

I couldn't agree more. I've been in this situation with clients' podcasts too. However, I do encourage the people involved in the podcast to not sound like they're reading. They usually accept my advice when I explain that we'll probably put people to sleep if we have three or four people around a table who sound like they're reading on the podcast.

Another of my clients in the financial industry takes a different approach. We make sure the questions I'm going to ask the guest are approved in advance. We then do several takes on the answers. Some guests get better as they warm up. Then I get a transcript of the interview and send it to the client. The legal and compliance people review it to make sure we're not going to get in hot water, and then I edit accordingly. Fortunately, we've become so adept at writing the questions that the lawyers rarely ask for edits.

If clients do request edits after the interview, you have to remind them that an audio file is not like a Word document. Rarely can you just clip out a word here and there. As Steve Lubetkin says, you really don't want to have to re-assemble the guests and re-record, but this could happen if the guests have ventured into territory that makes the lawyers nervous.

"The best interviews happen when I start off with questions that are merely guideposts, and let the conversation branch off naturally, with an easy flow."

CHAPTER 20
HOW TO WRITE A
PODCAST SCRIPT

DONNA PAPACOSTA

Why do I discourage scripting and reading a podcast? First, most people are pretty lousy script readers, as I've shared with you in the previous chapter.

Secondly, I think the best podcasts sound conversational rather than scripted. That being said, it is also true that if you can fake a conversational tone, you might do all right with a script. Faking entails inserting the occasional pause, *um, ah,* and even a flub or two. You should also be sure to vary the speed of your read.

Do you think I'm disingenuous recommending this faking tactic? Well, let's suppose you've read and agree with the advice and yet you have good reasons for writing and delivering a podcast script. For example, in some tightly wound organizations, every word uttered publicly must be approved, so it's deemed easier to get permission before recording rather than after. For some of us, this would be a last resort, but it may be necessary at times.

The main point is this: Please do not pick up a written document, which is crafted for the eye, and just read it cold. I can almost guarantee this won't work. You need material that is written for the *ear*.

WRITING FOR THE EAR

Sometimes people forget this simple truth: When we listen to what you are saying, it's not the same as reading what you have written. Copy that looks fine on the page or screen does not always sound right to the ear. Worse, it can be hard to understand. The listener does not have visual cues. In other words, there are no headlines, subheads, bulleted lists, sidebars or even paragraph breaks for him to see.

Take a typical written passage, adapted from copy on a municipal economic development website:

> With easy access to North America's third largest financial centre and a workforce armed with expert knowledge in a variety of fields, establishing your business in Hooterville will pay handsome dividends. Located at the epicenter of Canada's Golden Horseshoe, Hooterville, a dynamic community of 150,000 residents, is well within reach of major
>
> U.S. capital markets and nearly seven million potential consumers in southern Ontario. Coupled with a favourable Canadian tax environment, Hooterville makes perfect business sense.

This is fairly well written, except for the dangling participle in the first sentence. Let's now recast it for the spoken word:

> *Think about establishing your business in Hooterville. Here's why: Hooterville offers a talented workforce and easy access to North America's third-largest financial centre – Toronto. In fact, Hooterville is at the heart of Canada's Golden Horseshoe, with a dynamic community of 150,000 people. What's more, the town is close to major U.S. markets and almost seven million consumers in southern Ontario. Add a favourable Canadian tax environment, and Hooterville makes perfect business sense.*

Do you see the difference? Shorter sentences, simpler language, and a more direct tone.

WRITING TIPS

Here are a few tips for writing for the ear, whether your end product is a script, speech or podcast:

Use simple words, not complex ones. For example: "Use" rather than "utilize."

Shorten your sentences. If it requires a semicolon, it's probably too long.

Round all numbers. Say "nearly one million," not "989,320," unless there's a reason to use the exact figure.

Use the active voice, not passive. "Our team ran the webinar," not "The webinar was run by our team."

Use less formal language. For example, use contractions, as long as you can enunciate clearly. (*Won't* rather than *will not*.) Be sure listeners can hear the difference between your pronunciation of *can* and *can't*.

Give auditory guideposts. Say things like: "Let's talk about three ways to use social media to market an event. First..." Add transitions between each of your points, and a recap at the end, using your numbered list as a structure.

Don't be afraid of repetition. It's OK to repeat important information for emphasis.

"Copy that looks fine on the page or screen does not always sound right to the ear ... The listener has no visual cues ..."

Scripting introductions

I think it's fine to script the introduction of your podcast: "Hi, this is Mary Jones of Magnificent Consulting. Welcome to the Magnificent Podcast, where we talk about content marketing ..."

Write the intro down, then learn how to deliver it in a warm and confident way. You want to grab people's attention and draw them into the podcast.

When you *should* script the body of your podcast

When a script is called for, I recommend *lightly* scripting. Start with writing down the goal of this particular podcast episode. What are you trying to accomplish?

Explaining the new dental benefit to employees? Rallying the troops for next

quarter's sales campaign? Getting customers excited about your company's latest software release?

However, if you're a fiction writer, essayist or memoirist, and you are sharing your writings on your podcast, then by all means *read* your work to us. But please first learn how to read aloud. Vary the tempo. Slow down. Speed up. Introduce variety to your vocal skills; avoid a singsong cadence, which is an easy trap to fall into. Vary the loudness of your voice.

Be open about the fact that you're reading. Your fans don't mind at all; they *love* your written word!

CHAPTER 21
HOW TO READ A PODCAST SCRIPT NATURALLY

DONNA PAPACOSTA

As you know by now, although I'm not a big fan of reading in a podcast—unless you're deliberately sharing the passage of a book or article—I know podcast scripting is not uncommon in the real world.

Over the past few years, I have coached many people on how to record audio that doesn't come across as stilted or phony. This is more difficult than it sounds. As a podcaster, you want to sound like *you*, only better. As Oscar Wilde famously said, and my friend Mitch Joel likes to quote: "Be yourself; everyone else is already taken."

Whether you script or not, you don't want to sound like Ted Baxter, Ron Burgundy or any other fictional fool. So if you have decided, for whatever reason, to script your podcast, is it possible to sound natural? Yes, but it takes work.

Here are a few tips, based on my experience in scriptwriting, voice-over and podcasting.

Prepare. Do not read the script cold. Even trained actors can fumble a cold read.

Revise. Let's assume the script is well written, but still needs to be tweaked for the ear. Look for too-long passages, complex sentences, or phrases that can only be understood by the eye, not the ear. Revise them.

Mark up the script. If you can, print it out double-spaced and grab a red pen. Using whatever kind of marks you're comfortable with, add the following accents to your script: pauses, emphasized words, slower pace, faster pace.

Look up the pronunciation of any words you're not sure of and write them out phonetically if you need to. Be especially careful with names. Is it Bern-STEEN or Bern-STINE?

Practise by standing up and reading your script aloud. Don't sit in your chair and mumble through it!ß

Slow down. Yes, I can almost guarantee you're reading too fast. Remember: we are *listening* to your words, not reading them on the page.

Sound conversational. If you want to sound formal, because that's your style, go right ahead. If you want to sound conversational, be sure to pause occasionally and even make a mistake. Rare is the person who can speak for more than five minutes without an *um, ah, you know* or other misstep. Be human.

Double take. If you're not sure about how to approach a certain section, record more than one take. That's the beauty of recorded audio; you can edit later. If you pause between takes or drop a marker (if your software allows), or clap your hands loudly, which will leave a visible mark in your file, your edits will be easier to do.

This all might sound like a lot of work, and it can be. But think about the end result: do you want to move your audience or do you want to bore them? Text that sounds like it's being read is often dull and lifeless. If you take the time to revise and mark up your script and practise, practise, practise, you'll end up with a better piece of audio.

"If you take the time to revise and mark up your script and practise, practise, practise, you'll end up with a better piece of audio."

CHAPTER 22

THE FUTURE OF THE BUSINESS USE OF SOCIAL MEDIA (2024 - 2025)

THE DIGITAL LANDSCAPE is perpetually evolving, and as we look towards 2025, social media platforms will further cement their roles as titanic channels for business marketing and user engagement. Below, we explore the trajectories likely to shape YouTube, TikTok, Instagram, Twitter, Facebook Pages, podcasts, blogs, newsletters, and websites.

Businesses must stay attuned to these shifts to exploit emerging opportunities and mitigate impending challenges intrinsic to these platforms' growth and transformation.

YOUTUBE

YouTube will heighten its focus on long-form and educational content, live streaming, and the expansion of YouTube Shorts to challenge TikTok's short-form dominance. **Businesses should explore creating in-depth educational series related to their industry**, harness the immediacy of live events for product rollouts and utilise Shorts for quick, impactful messages.

TIKTOK

TikTok is set to broaden e-commerce functionality, accentuate content thematic around mental health and wellbeing, and pursue demographic diversification. **Businesses can leverage TikTok's shopping integration** to facilitate seamless purchase pathways and produce content resonating with a conscientious consumer base.

INSTAGRAM

Instagram's future hinges on potentiating influencer marketing, enhancing Stories and Reels usage, and rolling out small business features like shoppable posts. **Micro-influencer partnerships will be vital**, with authentic storytelling in ephemeral content formats to forge brand loyalty.

TWITTER/X

Twitter/X is expected to unveil novel creator monetization avenues and amplify Spaces to rival Clubhouse, whilst continuing its tenure as the bastion for real-time discourse. **Adopting Twitter/'Xs subscription model and engaging in Spaces discussions** will be strategic moves for businesses, helping them trade upon exclusivity and community interaction.

FACEBOOK PAGES

Enhancements to Facebook Pages are projected in business feature toolsets, e-commerce integration and community-led engagements via Groups and Events. **Businesses could amplify reach and service accessibility** by capitalising on these improvements to drive sales and foster community spirit.

PODCASTS

The podcasting domain will witness escalated competition, exclusive content proliferation, and advertising growth. **Creating or sponsoring niche podcasts** might yield lucrative dividends, while data-rich advertising approaches can sharpen targeting precision.

BLOGS

Connoisseur content with embedded multimedia elements will characterise the future blogosphere, with a pivot away from SEO-centricity towards authentically engaging niches. **Forging a unique narrative voice and integrating rich media** can help businesses resonate distinctively with audiences.

NEWSLETTERS

Curated, personable newsletter content with auditable engagement metrics will gain prominence. **Businesses ought to focus on circumspect content delivery and explore the convergence** of newsletters with existing content strategies for congruent brand experiences.

WEBSITES

Websites will spotlight mobile-first designs, personalized user journeys, and prioritise stringent security protocols. Businesses must **invest in responsive design paradigms, user interactivity**, and robust cybersecurity measures to safeguard user trust and compliance.

STRATEGY RECOMMENDATIONS

Navigating these imminent changes requires strategic agility and foresight. Businesses should consider:
- Investing in quality video and interactive content creation.
- Integrating direct-sales functionalities for social media channels.

- Partnership with influencers to harness their converging authority.
- Exploiting analytics for granular audience insights.
- Diversifying content forms to bolster engagement across platforms.

In essence, as social media platforms evolve to offer more interactive, personalized, and integrated shopping and content experiences, businesses must recalibrate strategies to exploit this new digital zeitgeist. These insights provide a foundational compass for steering your enterprise through the fluid dynamics of social media's future landscape.

The above information is not an exhaustive analysis but a strategic outline to guide businesses in aligning with the rapid transformation of social media platforms.

CHAPTER 23
CONCLUSION

THIS BOOK HAS CONSOLIDATED the crucial role that social media platforms hold in the modern business landscape. The platforms themselves have not only revolutionised communication but have also opened up a vista of opportunities for businesses to deepen their connection with consumers. The key takeaways presented in the preceding chapters underscore the multifaceted benefits that these digital tools offer when leveraged astutely.

Firstly, social media platforms have empowered businesses to engage with a global audience, fostering a sense of community and loyalty that transcends geographical barriers. The personalized interactions that these platforms facilitate have set a precedence; consumers now expect and value transparency and genuine dialogue with the businesses they patronise.

Secondly, the thoughtful deployment of analytic tools allows companies to glean insights into consumer behaviour with incredible precision. By analysing patterns and preferences, businesses can tailor their offerings with remarkable accuracy, thus meeting consumer needs more effectively—an essential feature of customer-centric service provision.

Thirdly, in the era of instant gratification, the real-time communication capabilities that social media grants cannot be overstated. A prompt response to a customer inquiry or complaint can be the differ-ence between a strengthened relationship and a lost customer. In these fleeting exchanges lies the potential for significant brand reinforcement or, conversely, reputational damage.

The call to engage consistently with consumers on social platforms is not one to be taken lightly. In an age where authenticity and trans-parency are not just preferred but demanded, it falls on businesses to uphold these values. Forging meaningful relationships with consumers through sincere engagement can create a stable foundation for a busi-ness, from which trust and loyalty can flourish.

Looking ahead, it is predicted that social media will continue to evolve and innovate, offering even more intricate tools for business-consumer engagement. The rise of AI and machine learning stands to refine the lashings of consumer data, giving birth to unprecedented personalisation in marketing efforts.

The migration towards video content also signals a fundamental transformation in the way businesses communicate with their audi-ence. It is within this dynamic and the immersive medium that the next chapter of corporate storytelling is set to unfold.

E-commerce capabilities within social media are another frontier likely to see considerable growth. Enabling consumers to transition seamlessly from browsing to purchasing within the platform itself will not only simplify the consumer experience but also open new channels for revenue generation.

Lastly, the reliance on social media for crisis management is expected to expand. Businesses can no longer afford to ignore the discourse on social media; instead, they must be poised to respond to consumer issues with agility and responsibility.

The key message here is one of adaptation; businesses must nimble-footedly respond to the advancements of social media or face the peril of obsolescence. Urgency underpins this message, for the landscape shifts not yearly, not monthly, but daily.

Thus, the imperative is clear: as businesses continue to chart their course through the digital age, their compass must unfalteringly point towards engagement, analysis, authenticity, and adaptability. Social media is not just a tool; it is the terrain upon which businesses must now build their empires.

This book has endeavoured to provide the map to that terrain. May you use it to chart a prosperous and innovative path for your own business in this exciting era of digital connection and commercial opportunity.

CHAPTER 24
MORE BOOKS BY LEE HOPKINS

* Podcasting: How to get started with podcasting in your organisation

* The 3 Es for business profit

* Brand identity: Why managing it is so important to your success

* 20 secrets you need to know about you, your business, and the Internet

* Accent & Tone of Voice

* Social media: Making social media work for your business (First edition)

* Social media: The new business communication landscape (First edition)

* Social media: Or how we stopped worrying and learned to love communication

* *Social media: Making social media work for your business (First edition)*

* *Social media: Measuring the impact and ROI of social media (First edition)*

* *Twitter mastery for business*

* *How to be your Possible Self*

* *Meningie Man*

CHAPTER 25
LEE'S CURRENT PODCASTS

* The Podcasting Podcast: How to get started with podcasting in your organisation.

* How to be your Possible Self

* BCR: Better communication results

* The ghost at the table